SpringerBriefs in Evolutionary Biology

SpringerBriefs present concise summaries of cutting-edge research and practical applications across a wide spectrum of fields. Featuring compact volumes of 50 to 125 pages, the series covers a range of content from professional to academic. Typical topics might include:

- A timely report of state-of-the art analytical techniques
- A bridge between new research results, as published in journal articles, and a contextual literature review
- A snapshot of a hot or emerging topic
- An in-depth case study
- A presentation of core concepts that students must understand in order to make independent contributions

Lize Hermógenes de Mendonça
Bernard Michaux • Malte C. Ebach

Biotectonics of Neotropical Transition Zones

 Springer

Lize Hermógenes de Mendonça (iD)
University of New South Wales
Sydney, NSW, Australia

Bernard Michaux (iD)
Kaukapakapa, New Zealand

Malte C. Ebach (iD)
University of New South Wales
Sydney, NSW, Australia

ISSN 2192-8134   ISSN 2192-8142   (electronic)
SpringerBriefs in Evolutionary Biology
ISBN 978-3-031-80161-7   ISBN 978-3-031-80162-4   (eBook)
https://doi.org/10.1007/978-3-031-80162-4

This Springer imprint is published by the registered company Springer Nature Switzerland AG
The registered company address is: Gewerbestrasse 11, 6330 Cham, Switzerland

If disposing of this product, please recycle the paper.

Preface

Biotic transition zones, and their likely drivers, are rarely discussed at the continental level, having far more prominence in single taxon, macro-ecological studies of smaller biogeographical provinces. With the onset of better geophysical techniques and faster computing power, as well as the plethora of large-scale databases and molecular phylogenetics, it is now possible to study biotic transition zones in context to plate tectonic processes.

This book shows how to apply a biotic and tectonic, or *biotectonic*, application to Neotropical biotic distributions, particularly at convergent plate margins. The Neotropics has a rich tectonic history and some of the greatest biodiversity in the world. Combined they can show whether tectonics is a potential driver of continental scale bioregionalisation, particularly transition zones.

This book reviews the history of transition zones and investigates the biota living on the Caribbean and South American plates, using comparative biogeographic analytical techniques. Chapter 1 reviews the history of transition zones in context to bioregionalisation. Chapter 2 reviews and analyses the tectonics and biota of the Caribbean plate and establishes whether the Caribbean (Antillean subregion) is in fact a transition zone. Chapter 3 investigates the South American plate, in particular the South American Transition zone (STZ), using molecular data, to determine whether the STZ is one or more transition zones, with different biotic and tectonic histories. Chapter 4 is a synthesis chapter, in which new developments in plate tectonic theory can determine which geological processes produce the different types of transition zones.

Sydney, NSW, Australia
Kaukapakapa, New Zealand
Sydney, NSW, Australia

Lize Hermógenes de Mendonça
Bernard Michaux
Malte C. Ebach

Competing Interests

Chapters 1, 2 and 3 are based on research for the doctoral thesis of Lize Hermógenes de Mendonça conducted at the University of New South Wales between 2019 and 2024. Chapter 1 was published in 2020 in an altered form in the *Biological Journal of the Linnean Society*. Lize Hermógenes de Mendonça was funded by Research Training Program Scholarship at the University of New South Wales. The authors have no conflicts of interest to declare that are relevant to the content of this work.

Contrasting Interests

Contents

Chapter 1
A History of Transition Zones in Biogeographical Classification

Abstract Transition zones are problematic in biogeographical classification as they represent artificial areas. A review of transition zones into existing biogeographical classifications shows conflicting area taxonomies. While many authors consider transition zones as overlap zones or areas of biotic mixing, only a few have considered excluding them from biogeographical classification altogether. One way of incorporating transition zones into a natural classification is by treating them as artefacts of geographically overlapping temporally disjunct areas. In doing so, geographically overlapping areas may occupy the same space but have different boundaries and histories. Temporally disjunct areas do form natural hierarchical classifications, as seen in the paleobiogeographical literature. A revision of each transition zone will determine whether they are artificial areas, areas within their own right or potentially geographically overlapping temporally disjunct regions.

1.1 Introduction

Biogeographic classification or bioregionalisation has its roots in biological taxonomy and Humboldtian geography (Ebach 2015), developing in the nineteenth century into historical biogeography and ecology (sensu Nelson 1978). Recently, with the advent of large distributional databases, geospatial techniques, and computing power, bioregionalisations have become quantifiable at a global scale (e.g. Kreft and Jetz 2010; Holt et al. 2013). Advances in quantification have also rekindled a long-standing debate as to what constitutes the boundary of a biogeographic area (e.g. kingdom, region, subregion, province). Are these boundaries clear-cut or fuzzy? Do biogeographic areas expand and/or contract through time? More importantly, what biological and Earth processes shape these areas, and how can we tell a historical area shaped by processes over millions of years from short-lived ecological zones? Of particular importance are known areas that do display characteristics of two large biogeographic areas. These transition zones share a high degree of overlap and have been noted by naturalists as far back as Lichtenstein (1827), Swainson (1835),

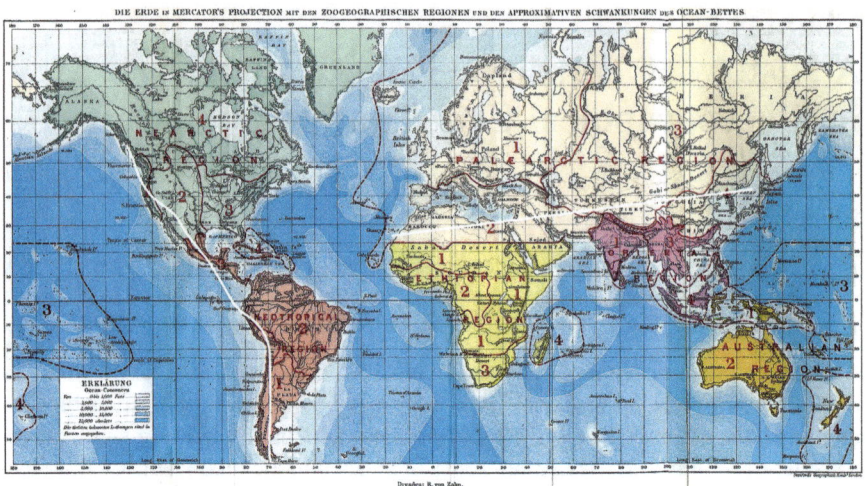

Fig. 1.1 'Map of the World, showing the Zoo-Geographical Regions and the contour of the Ocean-bed'. (Wallace 1876, frontispiece)

and Heilprin (1887), who described the natural mixing of organisms between large biogeographic regions, often referring to them as faunal 'crossing of paths', 'blending', 'transitions', or 'tracts'. The first to formalise transition zones as separate 'intermediate regions' was Charles Forsyth Major in his revision of Wallace's regions (Fig. 1.1) into five primary and three intermediate regions: (1) Holarctic Region (with the Circumpolar, Palearctic, and Nearctic subregions); (2) Oriental region; (3) Ethiopian region; (4) Australian region; (5) Neotropical region; (6) Mediterranean transitional region [Übergangsregion]: between the Holarctic, Ethiopian, and Oriental regions; (7) Austral-oriental transitional region [Übergangsregion]: between the Oriental and Australian regions; and (8) Nearctic-Neotropical transitional region [Übergangsregion]: between the Holarctic and Neotropical regions (Forsyth Major 1884: 113).[1] Forsyth Major was concerned that transition regions were being left off maps (e.g. in the case of Wallace 1876; Wallace and Thiselton-Dyer 1885) and deserved to be represented as natural or primary regions defined by fuzzy boundaries. Theodor Arldt made a similar observation, noting that where 'sharp boundaries are missing, transition zones [Übergangsgebiete] develop' (Arldt 1907: 21). Since 1884, transition zones have been treated as natural biogeographical regions or as areas in which the biota blend into one another. For example, compare Forsyth Major with Swainson:

---

[1] Heilprin (1887: 57) used the transition regions of Forsyth Major as (a) Tyrrhenian or Mediterranean transition region, (b) Sonoran or American transition region, and c. Papuan or Austro-Malaysian transition region.

> The situation with organogeographical regions is similar to that with the departments of systematics; depending on the point of view of the researcher concerned and depending on the point of view of knowledge, they will be more or less natural (Forsyth Major 1884: 102, our translation).[2]

> … but this transition, striking as it is, is not more conspicuous than that which may be traced from the zoology of Asia to that of America. It must be remembered, also, that each of these zoological provinces are connected with the rest at more than one point. The Asiatic blends into the European, both at its northern and at its western confines (Swainson 1835: 59).

Swainson did not make the same connection between systematics and area classification, namely, that zoogeographic regions are natural in the same way that taxonomic ranks were. In this sense, the concept of overlapping or transitional areas was impossible as the geography was fixed and taxic distributions were fluid. Any discussion about transition zones, regions, or areas would only make sense in light of natural biogeographic areas (sensu Forsyth Major 1884: 102).

Swainson's, and later Forsyth Major's, idea that biogeographic areas are natural and belong within a systematic area classification resonated with Wallace (1876), who insisted that such an area classification required a nomenclature to stem the already plethora of names for the same areas. Yet the bioregionalisation of biogeographic regions and realms was mostly a zoogeographic pursuit (Cox 2001), with many phytogeographers focusing on climate and the distribution of vegetation (i.e. biomes; Olson et al. 2001), rather than overlapping distributions of taxa (i.e. a biota; Morrone 2015). Early phytogeographers were more concerned with either geography (i.e. topography, climate and soils; Schouw 1823) or endemism (de Candolle 1820), meaning a shift in climate would bring about drastic changes in the distribution of vegetation. The result was smaller, more focused phytogeographic provinces compared to the larger zoogeographic regions, which were considered older and a result of climate as well as geological change (e.g. orogeny and basin formation).

Since Forsyth Major, both plant and animal bioregionalisations had undergone revision, although zoogeographic regions have the least change. For example, Wallace's regions and the three transition zones of Forsyth Major are still in use today, albeit in a modified form (Table 1.1).[3] Phytogeographical areas have undergone the most revisions since de Candolle (1820), with contributions and revisions offered by nineteenth-century Humboldtians, Schouw (1823), Grisebach (1866) and Engler (1899), and twentieth-century biologists adopting Wallace's realms but introduced smaller provinces (e.g. Takhtajan 1986, Good 1953; Udvardy 1975).

Modern biogeographers continue to use transition zones as discrete areas (e.g. Michaux 2010; Morrone 2015; Ferro and Morrone 2014; Ferro 2024). For example, transition zones are often referred to as areas that fit into a classification, such as

---

[2] Original reads 'Es verhält sich mit organogeographischen Regionen ähnlich wie mit den Abteilungen der Systematik; je nach dem Standpunkt des betreffenden Forschers und je nach dem Standpunkte des Wissens werden dieselben mehr oder weniger natÜrlich ausfallen'.

[3] Kreft and Jetz (2010: 2036) used the Nearctic–Neotropical, Sahara–Congo, Temperate–Tropical East Asia, which is equivalent to Forsyth Major's transition zones.

Table 1.1 Wallace's regions and Forsyth major transition zones

Authors	Region	Transition zones
Wallace (1876)	Palearctic	
	Ethiopian	
	Oriental	
	Australian	
	Neotropical	
	Nearctic	
Forsyth Major (1884)		Mediterranean
		Austro-oriental
		Nearctic-Neotropical

Wallacea in the Oriental realm (Dickerson et al. 1928; see Parenti and Ebach 2009), rather than an overlap zone where one area ends and another begins. If placed into a systematic classification, how would transition zones relate to one another and to the larger regions to which they belong? The aim of this chapter is to identify transition zones, as natural areas with a shared biotic history or as artificial areas of overlap, within a biogeographical classification (see below).

1.2   Biogeographical Classification: Area Taxonomy and Natural Area Classification

Biogeographical classification has a long history, particularly within the German and French literature (Ebach 2017). The first biogeographical classifications were based on traditional geographical boundaries, such as the Old and New Worlds (Zimmermann 1778–1783), climatic provinces (Fabricius 1778), latitude (Giraud Soulavie 1780), vegetation (Humboldt and Bonpland 1807), natural areas based on endemism (de Candolle 1820; Prichard 1826; Swainson 1835), endemism and vegetation (Schouw 1823), oceanic isotherms (Forbes 1846), natural areas based on animal forms (Schmarda 1853), and life zones (Merriam 1892). From this vast literature, two types of classification prevailed: a taxonomic classification, based on the numbers of endemic taxa, and a vegetational (non-taxonomic) classification, based on vegetation and climate. These two types of classification dominated twentieth-century biogeography, either individually or in combination (Good 1953; Udvardy 1975, Takhtajan 1986). Later, the terms ecoregion (Olson and Dinerstein 1998) and ecozone (Olson et al. 2001) appeared, in line with the ecological-based classifications of Udvardy (1975). These two types of classifications, one based on endemism and vertebrate distributions and the other on ecosystems and vegetation, are a holdover from the nineteenth-century debate on what constitutes 'natural'.

1.2.1 Natural Classification

In biological taxonomy, a natural classification in taxonomy is a hierarchical ordering of groups called taxa that are presumed to be more closely related to each other than to any other taxon. In other words, taxa are presumed to be natural until shown to be otherwise. Natural classification has been the goal of taxonomy since the time of early naturalists, such as Carl Linnaeus, Antoine Laurent Jussieu, and Augustin de Candolle; however, a method for establishing natural taxa was not formalised until the mid-twentieth century (Rosen 1978; Nelson and Platnick 1981).

The idea of a natural area classification was debated throughout the nineteenth and early twentieth centuries. The Humboldtian tradition held that vegetations were natural given that the same abiotic factors determined the same type of vegetation (sensu Schouw 1823). For instance, a warm and wet climate and well-drained soil would produce a rainforest, regardless of whether it was in New Guinea, Africa, or South America. In contrast, the taxonomic tradition held that species were endemic to a particular area (sensu de Candolle 1820) and that taxonomy and distribution alone were sufficient to determine natural areas. A hybrid application, in which larger areas were deemed older and thereby historical and smaller areas younger and thereby ecological (see Nelson 1978), was adopted largely by the Humboldtians in their phytogeographic classifications. While many assume that the taxonomic and Humboldtian traditions have been successfully combined in an ecoregion classification (Olson et al. 2001), little attention has been paid to whether these ecoregions are in fact natural and analogous to a natural classification in biological taxonomy.

The problem is that abiotic factors may create similar environments; the organisms that evolve in these environments are not the same. For example, the rainforests of New Guinea, Africa, and South America contain very different taxonomic groups that share no common history. In this sense, New Guinean and African rainforest biota are similar in form but not in taxonomic classification. Moreover, closely related biota may exist in very different ecological areas. The biota in the Eastern Desert province in the Australian region are more closely related to temperate provinces along the southeastern margin of Australia than they are to any other desert province. It is clear that ecological classifications of the Humboldtian tradition are not natural as they do not share a common history. If we were to base a natural area classification on the successful natural classification in biological taxonomy, we would need to equate taxa with areas, which leads us to the next problem: what do we mean by areas?

1.2.2 Areas as Taxa

In biogeography, areas are often understood to be one of three things, taxic (biotic) distribution, geography, or biomes. Biomes are traditionally thought to be an area dominated by the same vegetation type and climate, akin to an ecoregion. The definition of biomes has subsequently changed and is largely based on climate (e.g. desert biomes; Crisp et al. 2004). Biomes generally are understood to be non-historical, but attempts have been made to incorporate biomes with historical aspects, such as species distributions (see Crisp et al. 2011). Areas as geographical features, such as plains, plateaus, and ocean basins, are a common concept in paleobiogeography, as our knowledge of taxic distributions is poor given the incompleteness of the fossil record. For example, if a single trilobite specimen is found in a shallow marine basin, such as an epicontinental sea, the whole basin would be designated as the area of distribution. Taxic distribution, mostly species distributions, is the focus of biogeographical research, particularly in geospatial studies (see Holt et al. 2013; Kreft and Jetz 2010). For the purposes of a natural classification, an area would be designated as a biotic distribution, that is, the overlapping distribution of all taxa that are endemic to an area. How those areas are determined is similar to how taxa are proposed in biological taxonomy: by the practitioner who has spent years looking at specimens. This is not to say that there is not a dearth of computation methods (see Crisci et al. 2003), but many of these methods are models that are based on inferences made by practitioners who have spent years looking at the distribution of taxa. What is important is not which models are the best for defining biotic areas, but how we test whether these areas are real (i.e. natural) or not. Given that a natural classification aims to find whether taxa are monophyletic, that is, that some taxa are more closely related to each other than they are to any other taxon, the same can be said for a natural area classification. A natural area classification or area taxonomy (Ebach and Michaux 2017) aims to find an area monophyly in order to test whether existing named areas are monophyletic (i.e. natural) or not. A monophyletic area is equivalent to a natural classification (Williams and Ebach 2020).

1.2.3   Area Taxonomy

An area taxonomy is analogous to a biological taxonomy, in that areas, like taxa, form a natural hierarchical classification (Ebach and Michaux 2017). Area taxonomy also includes an area nomenclature (Ebach et al. 2008), which formalises the naming of areas to prevent area duplication. Presently, there are several area taxonomies that function as a whole under the same International Code of Area Nomenclature (ICAN; Ebach et al. 2008). These area taxonomies cover the Australian region (Ebach et al. 2013; 2015, 2020), the Neotropical region (Morrone 2017), the Andean region (Morrone 2018), and the world (Morrone 2015). The area taxonomies of Morrone (2015, fig. 1; Table 1.2, Fig. 1.2), as well as the ICAN, are used in this book.

While the concept of an area taxonomy is over 200 years old, its formulation as a distinct field of research is relatively new. Given this, many of the problems faced by biological taxonomists have not yet been discussed in area taxonomy, such as monotypic areas, revisions of well-known areas, and transition zones.

Table 1.2 Area taxonomy of Morrone (2015)

Kingdom	Region	Transition zones
Holarctic	Nearctic	
	Palearctic	
Holotropical	Neotropical	
	Ethiopian	
	Oriental	
Austral	Cape	
	Andean	
	Australian	
	Antarctic	
		Mexican
		Saharo-Arabian
		Chinese
		Indo-Malayan
		South American

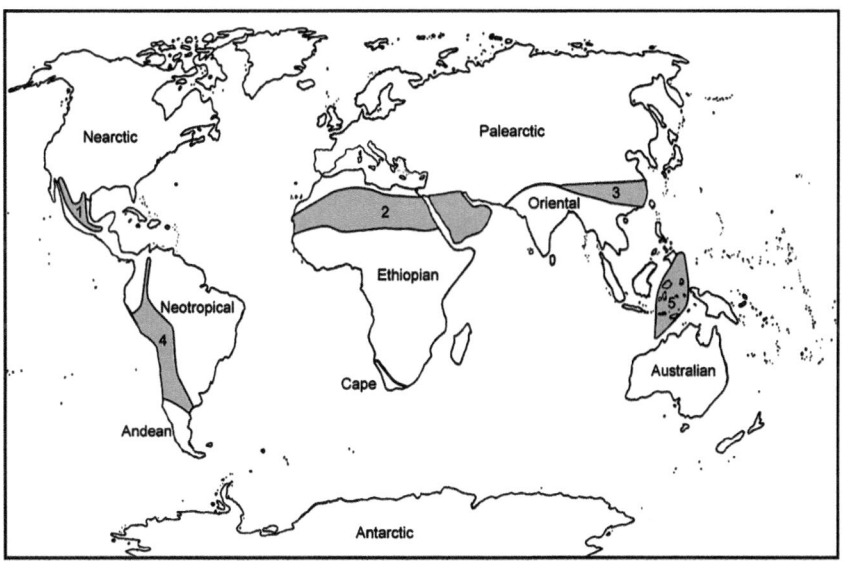

Fig. 1.2 World biogeographical regionalisation showing regions (white) and transition zones (grey). Transition zones: 1 = Mexican TZ, 2 = Saharo-Arabian TZ, 3 = Chinese TZ, 4 = South American TZ, 5 = Indo-Malayan TZ (Wallacea). (Morrone 2015, Fig. 1)

Monotypic taxa are problematic in taxonomy as they are aphyletic, in that we do not know if they are monophyletic or not (Ebach and Williams 2010). Monotypic taxa have the potential to create aphyletic classifications, in that many are often not included in nested taxa (i.e. clades) and are often left in a rank of their own, creating further monotypic taxa. The revision of well-known areas has the potential to create nomenclatural problems. For example, in order to make the aphyletic taxon Dinosauria monophyletic, we would need to include the birds, which have the nomenclatural implication of making the Dinosauria synonymous with the older name Aves. There have been attempts to resist revising such groups due to the risk of losing a popular name (see Brummitt 2002).

Transition zones, or areas of regional overlap, are often treated as single areas (e.g. Dickerson et al. 1928; Burbidge 1960), thereby creating hybrid areas. For example, Dickerson et al. (1928) created the area Wallacea referring to the triangular area that exists between Wallace's and Weber's line. Wallacea contains both biota from the Oriental and Australian regions, making it a hybrid, namely.

Holotropical kingdom.

>   Oriental region.

>   Wallacean subregion.

Austral kingdom.

>   Australian region.

>   Wallacean subregion.

If Wallacea is treated as a subregion, then we have the problem that it falls equally in two separate regions with two very different histories *at the same time*. Doing so is problematic as the same region cannot be represented twice in a classification. If it is represented once, it will cause aphyly to occur. Presently, there are five recognised transition zones (Morrone 2015), meaning ten or more possible places within a classification. What, then, are transition zones, and how do we resolve them within a natural classification?

1.3   Transition Zones

1.3.1   Definition of a Transition Zone

The definition of 'transition zone' has remained consistent when it was first coined by Clinton Hart Merriam in 1892:

> The Transition Zone, as its name indicates, is a zone of overlapping of Boreal and Sonoran types (Merriam 1892: 31).

> Where the sharp boundaries are missing, transition zones [Übergangsgebiete] develop, for instance in America the so-called Sonorian or Meso-Columbian region are representative. In Asia we see a mixed fauna in China and, in the Indo-Australian islands from Celebes to the Moluccas and Timor and for many animal forms as far as the eastern Melanesia. These transition zones are comparatively small and are found along the mountainous and deserts borders of regions (Arldt 1907: 21).

> …the overlap of faunal elements with subtractions in both directions (Darlington 1957: 455).

...a zone of intense transition and superposition of faunas (Halffter 1962: 2).

... the transition among biogeographical regions is a phenomenon that starts when a possibility of biotic exchange among different regions (two at least), is established (Palestrini and Zunino 1986: 1).

... the boundaries between biogeographical regions, representing areas of biotic overlap, which is promoted by historical and ecological changes that allow the mixture of taxa belonging to different biotic components (Morrone 2006: 469).

... a geographical area of overlap, with a gradient of replacement and partial segregation between biotic components (sets of taxa that share a similar geographical distribution as a product of a common history) (Ferro and Morrone 2014: 1).

Definitions of transition zones appear to come with a caveat on their formation, either as a result of some process (e.g. biotic exchange) or with a particular example ('overlapping of Boreal and Sonoran types'). In order not to speculate on the processes that create transition zones (e.g. they may be different in every case), we simply define a transition zone as *an area of biotic overlap*. Our concern is that transition zones, like aphyletic taxa, are merely artificial constructs created by well-intended biogeographers. Given that transition zones are artificial (i.e. man-made) and therefore not a part of a natural classification, we need to understand what purpose they serve in biogeography other than identifying areas of biotic overlap.

For the purposes of identification, transition zones are both hazy and varied, depending on who identifies them. The Chinese Transition Zone (CTZ), for example, is figured in Stegmann (1930, taf. 15) as confined to the northern side of the Amur River; in Müller (1979, fig. 27), as an area covering much of southern China and Taiwan, and in Palestrini et al. (1987, fig. 2), it includes northern and western China as well as Korea and parts of Japan. The logic behind the notion that two regions will never share a sharp boundary would translate to the borders of transition zones (sensu Major Forsyth 1884). Perhaps this is why the borders of transition zones, such as CTZ, are often disputed. If individual transition zones themselves defy definition, why do we need them, particularly when, as Morrone (2015) noted 'instead of sharp boundaries, more flexible transition zones seem a better solution' (Morrone 2015: 83)? The answer is in that transition zones are areas of biogeographic interest as they do contain a high degree of biotic diversity and intermixing of biotas. Rather than ignore transition zones within an area taxonomy, it would be prudent to accept them either as areas in their own right (as a hybrid area classed in whichever region has the largest proportion of its biota) or as two separate regions that do not share the same boundaries.

1.3.2 Transition Zones as Hybrids

The trend for treating transition zones as hybrids (sensu Ferro and Morrone 2014; Ferro 2024) means that a decision must be made as to where it occurs in a biogeographical classification. For example, the Mexican Transition Zone (MTZ), a mixture of the Nearctic and Neotropical biota, is grouped in the Neotropical region (Morrone 2017), even though Morrone (2015) insisted that it also belongs to the Nearctic region.

A statistical study may indicate which biota, the Neotropical or Nearctic, dominates the MTZ. In order to do this, a simple method could be implemented, such as a count of all species within a transition zone and then dividing these into regions. The problem with such an approach is that (1) not all species are known from a region and (2) some higher taxa may dominate the described biota more than others. For example, sampling of invertebrate taxa in the interior Australian provinces is quite poor, whereas plant and vertebrate sampling is high. Will vertebrate taxa, for instance, give us the same percentile as plant or invertebrate taxa? A study by Di Virgilio et al. (2013) has identified a biotic break between two Australian provinces. The biotic break consisted of a 20 km rain shadow due to topography, resulting in an unusually dry area that does not seem to affect mammals but does have a pronounced impact on plant taxa (i.e. 40% turnover). If we discover that plant taxa, rather than bird and mammal taxa, are more affected by transition zones, could we select exemplar taxa to determine biotic dominance? The problem is that it turns into a numbers game, meaning that the result would depend completely on the choice of taxa made. Geospatial models could identify transition zones as unique clusters of taxa and therefore validate them as areas in their own right. Geospatial analysis would not be able to identify transition zones as areas of overlap since overlap does not cluster (e.g. Rosen 1988; Laffan et al. 2010). If searching for transition zones, clustering models can test if existing transition zones are in fact real unique and endemic areas. Modern authors have sought to formalise transition zones as areas within a single biogeographical classification (see Holt et al. 2013; Kreft and Jetz 2010 and below). To our knowledge, only one study by Smith (1984) had modified boundaries in order to minimise distributional overlaps between units. This is a departure from past practices in which boundaries have usually been associated with the most prominent 'breaks' in distributional patterns. The wisdom of this change might be questioned but can be defended on the grounds that it produces both a less redundant classification and one of greater internal order (Smith 1984: 461). Another approach is to treat transition zones as intermediaries that lie awkwardly within a classification (Major Forsyth 1884; Morrone 2017).

1.3.3   Transition Zones as Temporal Areas of Overlap

An alternative to transition zones is to treat biotic overlap as a temporal phenomenon. Time is rarely considered in the area taxonomy of recent biota. While biogeographers use molecular clocks in their biogeographic dispersion models, they rarely are used to determine temporal areas. In paleobiogeography, dating the ages of biotic areas is vital, as biota diverge and become extinct. A study by Dowding and Ebach (2018) has shown that regions and subregions change shape, converge, and diverge over time. The practice of proposing areas in the age of tectonics (from the

1960s) was led by palaeontologist Art Boucot, who proposed several new regions based on new paleogeographic maps and fossil distributions (Boucot et al. 1969). Although transition zones were not discussed, it was evident that every region and subregion had a different age; that is, the areas were not contemporaneous. In the Devonian, the Malvinokaffric, for example, still had a Silurian biota, while, at the same time, other regions had a more 'modern' Devonian biota. Given this, where these temporally disjunct areas met, there were transition zones that contained a mixture of Silurian and Devonian taxa (e.g. Melbourne basin, Dowding and Ebach 2016).

A study by King and Ebach (2017) found that by dating living taxa using molecular clocks, areas could be divided temporally to create two geographically overlapping but temporally disjunct areas. The study split biotic areas into Palaeogene and Neogene counterparts based on dated nodes in molecular clocks. The study proposed that if the areas were temporally split, they form monophyletic taxa based on age rather than on geography. The analysis by King and Ebach (2017) confirmed this hypothesis, namely, that Palaeogene and Neogene biotic areas were monophyletic, raising the point of the relevance of transition zones. If regions are temporally disjunct, then geographical overlap is simply an artefact, an area identified solely because it has a mixed biota. In this sense, transition zones are atemporal geographical analogues (sensu Humphries and Ebach 2004). In order to see past this atemporal overlap, the areas need to be mapped onto two or more maps in order to show the distribution in relation to past and/or present barriers. In this sense, transition zones will not appear as artefacts. To summarise, two areas of different ages can occupy the same space but have different boundaries. This geographic overlap forms an artefact that is often misunderstood as a transition zone.

1.3.4 Towards a Temporally Sensitive Area Classification

Presently, Neogene and Quaternary area taxonomy is treated as atemporal, which seems unusual given that paleobiogeographic studies have consistently shown that most areas are temporally disjunct. Palaeontologists may have a poor fossil record to contend with, but each fossil has an age. The opposite is true in neontology, but recent developments on molecular clocks have made it possible to date clades, that is, taxa, meaning that areas can be temporally sliced into two distinct biotic areas. In such temporally sensitive area taxonomy, transition zones will not appear on static, that is, atemporal maps because the age ranges of the areas can be shown in a series of maps representing different time periods or epochs. This is presently practised in paleobiogeography and can be easily adopted in biogeographic studies of recent taxa using molecular clocks (i.e. King and Ebach 2017). But what of clearcut barriers? If areas do have a similar or exact age range, then would overlap occur? In the Australian example, in which plant taxa have a 40% species turnover, the boundary represents a significant barrier. A barrier that confines 40% of taxa is not at the regional scale (Simpson 1977 proposed 70% or above) but represents a provincial barrier. The rest of the plant taxa—60% of species—exist in both provinces. Is this an area of transition? Presently, all transition zones are along or near convergent plate margins, where two areas are being thrust into one another. If we were to imagine a diverging plate margin, the reverse would happen, and at some point, a transition zone would form, one that may be similar to the ones we see presently. If this is the case, then we would have a high species turnover rate, and a divergent margin would create a barrier early in its development (e.g. a basin such as the Red Sea and Rift Valley in Africa), thereby isolating two biota that go on to evolve independently. There also will be taxa that may not respond to the barrier and thereby occur in two areas at the same time, as in our Australian example. Such taxa will create an area of geographical overlap and create the transition zone artefact. In this sense, transition zones cannot tell us about the process of biotic mixing (e.g. whether it is convergent or divergent). All it tells us is that there is a temporal disjunction between the biota; that is, there are two temporally distinct areas occupying the same area. A review of each of the existing five transition zones and their tectonic histories may help us understand whether these are truly artefacts or real areas that have a distinct biota.

1.4   Transition Zones: A Review

All five transition zones as summaries by Morrone (2015, fig. 1; Fig. 1.2) are reviewed below in order to determine their history in biogeographic classification. Each transition zone will be considered in terms of area taxonomy.

1.4.1   Chinese Transition Zone (CTZ)

The first mention of a transition zone within the northeastern part of the Palearctic was made by Alfred Russel Wallace:

> The fourth, or Manchurian sub-region, consists of Japan and North China with the lower valley of the Amoor [Amur]; and it should probably be extended westward in a narrow strip along the Himalayas, embracing about 1,000 or 2,000 feet of vertical distance below the upper limit of trees, till it meets an eastern extension of the Mediterranean sub-region a little beyond Simla. These extensions are necessary to avoid passing from the Oriental region, which is essentially tropical, directly to the Siberian sub-region, which has an extreme northern character; whereas the Mediterranean and Manchurian sub-regions are more temperate in climate. It will be found that between the upper limit of most of the typical Oriental groups and the Tibetan or Siberian fauna, there is a zone in which many forms occur common to temperate China. This is especially the case among the pheasants and finches (Wallace 1876: 72).

Wallace considered the limits of the Manchurian subregion as 'not very well defined'; however, it includes Japan and 'the Corea and Manchuria to the Amoor river and to the lower slopes of the Khingan and Peling mountains; and China to the Nanlin mountains south of the Yang-tse-kiang'. On the coast of China, the dividing line between it and the Oriental region seems to be somewhere about Fo-chow, but as there is here no natural barrier, a great intermingling of northern and southern forms takes place (Wallace 1876: 220–221). The 'intermingling' included the Coleoptera, 'with some naturalists a matter of doubt whether the southern and best known portion of the islands should not be joined to the Oriental region' (Wallace 1876: 228). Forsyth Major, however, concluded that the 'Manchurian sub-region, which also includes Japan, is made by Wallace a Palaearctic region, but has just as many affinities with the oriental, and is therefore more correctly understood as a transitional region between the two' (Forsyth Major 1884: 110). Theodor Arldt too noted 'a mixed fauna in China' (Arldt 1907: 21), yet neither he, Wallace, or Forsyth Major gave a clear indication, either in the text or on any map, where the transition zone was located. For example, Stegmann, working on bird distributions, noted that the 'actual transition must be seen to the north and west, in the Amurland, in northern Manchuria, and in Transbaikalia. Since it is of the utmost importance to establish the transition zone between these two so different parts of the Palearctic, I undertook two trips to southern Siberia in order to get to know at least a part of the areas in question. My main task was to study the animal avifauna of these areas with regard to animal distribution (borderline) for the Chinese-Japanese sub-area characteristic forms' (Stegmann 1930: 390).

Entomologists, however, designate Wallace's entire Manchurian subregion within the Chinese Transition Zone (CTZ) based on some groups of Scarabaeoidea (Coleoptera) distributions (Palestrini et al. 1985; Palestrini and Zunino 1986) and sweat bees (Apoidea: Halictidae: Halictinae; Ebmer 1998: 371; see also Kreft and Jetz 2010). Müller (1972, 1979), based on the synthesis of data relating to a large number of groups, indicated that the CTZ[4] is located in the area between the Yang-Tze-kiang and the 21 parallel, including also Taiwan. However, Palestrini et al. (1985) defined a more extensive area to the North. They define it as a band between the meridians 90 and 140 and, roughly, between the Tropic of Cancer and a line, in the first instance parallel to the Reinig line, which is from 50 ° N at the eastern-most longitude gradually changing to 35° N at the most western point. In other taxa, such as the brush-footed butterflies (Nymphalidae: Satyrinae) the 'Sino-Himalayan sub-region is a widely extended transition zone between the Palaearctic and Oriental region' (Deodati et al. 2009: 107), extending well beyond China. In a rare paleontological example, Fortelius and Zhang (2006: 131) considered that North Chinese land mammal faunas 'cyclically alternate and mingle in a transition zone' in the Miocene but do not indicate the exact location.

The CTZ features predominantly in the entomological literature and is completely absent in phytogeographic and macro-ecological studies (e.g. Smith 1984; Udvardy 1975; Olson et al. 2001), although some authors accept that an area of

---

[4] The term Chinese Transition Zone may have been coined separately by Palestrini et al. (1985) and Russel et al. (1951), the latter defining the CTZ as part of seven 'Culture Worlds'.

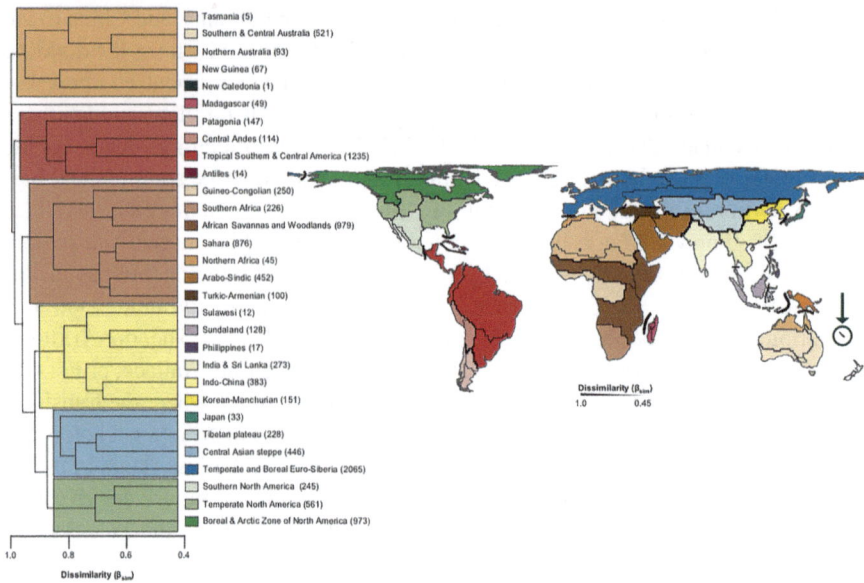

Fig. 1.3 The six major biogeographical divisions are shown in the dendrogram. Orange = Australasian, red = Neotropical, brown = African, yellow = Oriental, blue = Palaearctic, green = Nearctic (Kreft and Jetz 2010, Fig. 9)

transition exists, 'A different obstacle to the drawing of a faunal boundary is presented in South China, where the transition from the temperate Palearctic to the tropical Oriental Region is by uniform gradation, the result of long-continued progressive intermixture of the Oriental and Palearctic faunas' (Schmidt 1954: 329). Recent geospatial studies of Kreft and Jetz (2010; Fig. 1.3) using mammal distributions found an Eastern Palaearctic–Oriental transition zone, which corresponds generally to Wallace's Manucharian subregion. A similar analysis using vertebrates suggests 'the Sino-Himalayan cluster, its inclusion in the Palearctic region was a truly borderline one, because it could be almost equally well incorporated in the Indo-Malaysian region, although its subregional status in a global context is less questionable' (Procheş and Ramdhani 2012: 267). Holt et al. (2013), also using vertebrates, identified twenty distinct zoogeographic regions, which were grouped into eleven realms, including the Sino-Japanese realm that was bounded between the Palearctic and oriental regions. In a reply, Kreft and Jetz (2013: 343) pointed out that the new Sino-Japanese realm is in fact a 'well known' transition zone and a region of 'complex faunistic interchange', without providing any justification.[5] The bioregionalisation of the freshwater zoogeographical areas of mainland China (Huang et al. 2020) also recognises the Chinese transition zone as an area of overlap between the Palearctic and the Oriental regions, as indicated by Morrone (2015).

---

[5] Kreft and Jetz (2013) cite *History of the fauna of Latin America* (Simpson 1950), which makes no mention of East Asia.

Table 1.3 Placement of the CTZ in existing area taxonomies

Authors	Realm/Kingdom	Region	Subregion
Kreft and Jetz (2010)	–	Oriental	Korean-Manchurian
Procheş and Ramdhani (2012)	–	Palearctic	Sino-Himalayan
Holt et al. (2013)	Sino-Japanese	–	–

The CTZ could be considered a separate area as clustering models have found it on separate occasions as a separate 'Sino-Japanese realm' (Holt et al. 2013), as the 'Korean-Manchurian' within the Oriental region (Kreft and Jetz 2010) and as the 'Sino-Himalayan subregion' within the Palearctic (Procheş and Ramdhani 2012; Table 1.3). Based on these studies, it would be prudent to treat the Chinese Transition Zone as a potential natural area.

1.4.2   Saharo-Arabian Transition Zone

The Saharo-Arabian transition zone[6] comprises the Sahara Desert and the Arabian Peninsula and corresponds to the boundary between the Palearctic and Ethiopian regions (Müller 1986; Kreft and Jetz, 2013; Table 1.4). Müller (1986) stated that 'during the glacial stage extratropical, Mediterranean species migrated to the southern edge of the present-day Sahara. The Saharo-Arabian fauna and flora elements were pushed back to the hyper arid central areas of the Sahara. Tropical savanna species in the western part of the Sahara were able to advance as far as the Moroccan Sou Valley during the post-glacial periods. After the expansion of the central arid plains about 3000 years B.C., the Palaeartic elements were pushed northwards, i.e., to higher elevations within the Saharan mountains' (Müller 1986: 20). Kreft and Jetz (2013) considered the Saharo-Arabian as an impoverished set of Afrotropical lineages, but with strong influences from the neighbouring Palearctic and Oriental faunas. A similar argument by Cox (2001) stated that the Sahara and northern Africa should be regarded as impoverished parts of the African realm (Fig. 1.4). Other authors such as Smith (1984) placed the entire Sahara and the northern part of the Arabian peninsular into the Mediterranean subregion.

Schmidt proposed the Eremian Province that separates the Ethiopian subregion from the Palearctic and the Oriental 'by a chain of deserts that serve as highways for the dispersal of desert-adapted animals, and this has made for faunal interchange between the African and Eurasian continents. The establishment of an Eremian Province of the Palearctic recognises this as one of the major historical transitional areas' (Schmidt 1954: 229). Schmidt's Eremian is problematic, as it resembles a desert biome and not an area defined by taxic distributions. A similar claim for a desert realm was made by Holt et al. (2013), using geospatial clustering models,

---

[6] In phytogeography the same region is sometimes referred to as the Saharo-Sindian region (Wickens 1976; Brenan 1978, Kreft and Jetz 2010).

Table 1.4 Placement of the SATZ in existing area taxonomies

Authors	Realm/Kingdom	Region	Subregion
Schmidt (1954)	Arctogaean	Palearctic	Mediterranean
Smith (1983)	–	Afro-Tethyan	Mediterranean
Cox (2001)	African	–	–
Kreft and Jetz (2010)	–	Oriental	Korean-Manchurian
Procheş and Ramdhani (2012)	–	Palearctic	Sahero-Arabian
Holt et al. (2013)	Saharo-Arabian	–	–

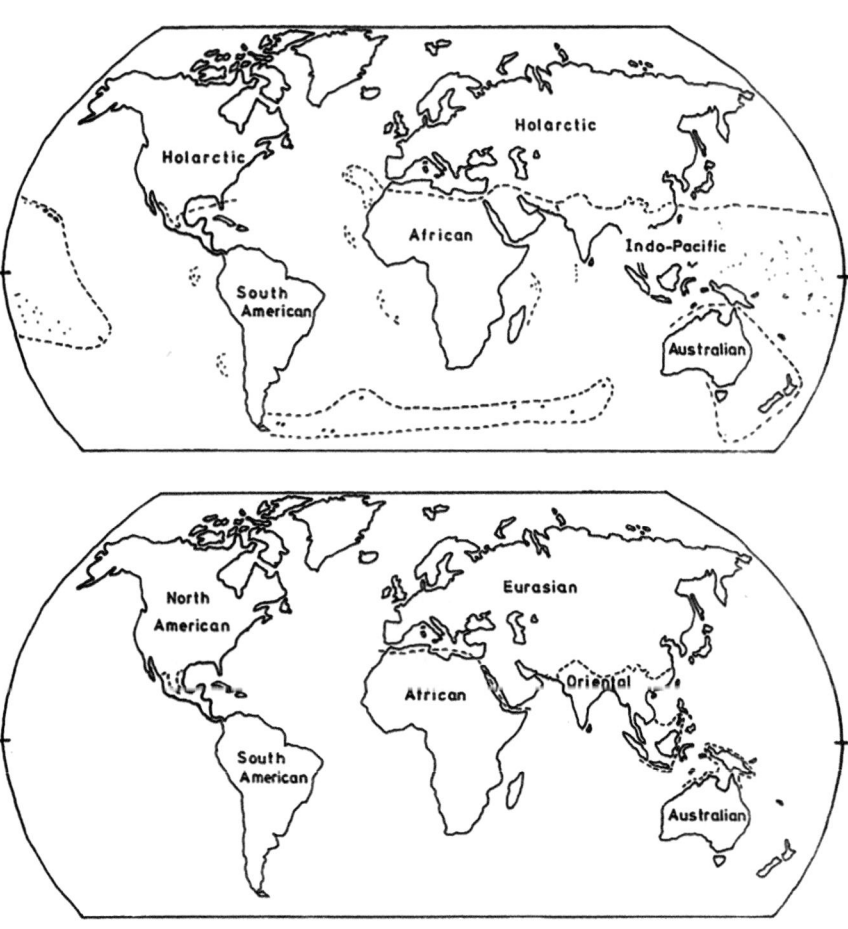

Fig. 1.4 Floral kingdoms (above) and mammal zoogeographic regions (below) of Cox (2001, Fig. 3)

who classified the Saharo-Arabian as a new realm that is 'intermediate between the Afrotropical and Sino-Japanese realms' (Holt et al. 2013: 75). The authors do not indicate what is meant by 'intermediate' and whether it simply describes the biota or whether it identifies the new realm as a transition zone.

Existing area taxonomies treat the Saharo-Arabian as a separate region, one that has appeared to cluster in the three similarity analyses. Those who have not classified it as a separate area have included it in either the Ethiopian or Palearctic region. The Saharo-Arabian transition zone is herein considered to be a potential natural area.

1.4.3 Indo-Malayan Transition Zone

The Indo-Malayan, Indonesian, or Wallace's transition zone (Darlington 1957; Müller 1986; Kreft and Jetz, 2013) corresponds to the boundary between the Oriental and Australian regions. Müller (1986) discussed its boundaries and gave examples of Oriental and Australian taxa with overlapping distributions in this transition zone (Morrone 2015; Table 1.5). An earlier treatment by Schmidt (1954) placed the Celebesian [i.e. Indo-Malayan Transition Zone] province into the Oriental region, contradicting Wallace (1876), who placed it in the Australian region. Unfortunately, Schmidt (1954) gave no justification for this arrangement. Earlier, Mayr (1944) attempted to draw a closure to the debate over a possible transition zone, proposed earlier by Salomon Müller (1846), Pelseneer (1904), Dickerson et al. (1928), and Rensch (1930) [Zwischengebiet], by stating that it 'is the home of four different faunas. It is self-evident that the formal recognition of a zoogeographic region of such heterogeneity is neither practical nor scientifically defensible' (Mayr 1944: 10). Rather, Mayr (1944) suggested Pelseneer's Weber's line as 'the best possible border line between the Oriental and the Australo-Papuan Regions' (Mayr 1944: 10). Simpson (1977) rejected Mayr's suggestion, 'I propose a radical but simple solution: let us stop playing the game. If we like (on the whole I do), let us keep the Oriental Region bounded by the Sunda Shelf and Huxley's Line and the Australian Region, bounded by the Sahul Shelf and Lydekker's Line, but let us not assign the intervening islands to any region, subregion, transitional or intermediate zone, or the like' (Simpson 1977: 118). This of course leaves a significant gap in any

Table 1.5 Placement of the Indo-Malayan TZ (Wallacea) in existing area taxonomies

Authors	Realm/Kingdom	Region	Subregion
Schmidt (1954)	Arctogaean	Palearctic	Oriental
Smith (1983)		Island	Australian
Cox (2001)	Oriental/indo-Pacific	–	–
Kreft and Jetz (2010)	–	Oriental	Sulawesi
Proches and Ramdhani (2012)	–	Wallacean	–
Holt et al. (2013)	Oriental	–	–

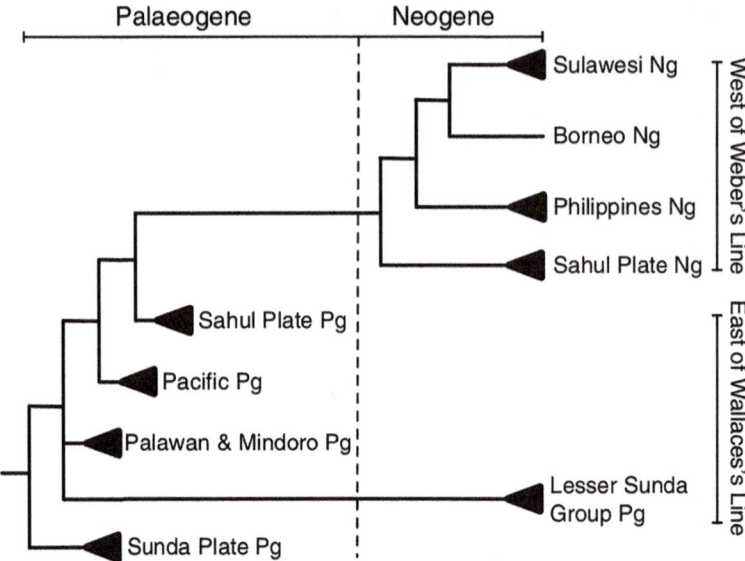

Fig. 1.5 Simplified general area cladogram representing groupings of biotic areas. Areas to the right of the dotted line are Neogene delineated by Weber's line, whereas areas to the left are Palaeogene and delineated by Wallace's line (King and Ebach 2017, Fig. 5)

biogeographical classification. Surely, the areas excluded by Simpson (Sulawesi, Mollucas, Flores, Timor) must be more closely related to an area in one region than it is to another elsewhere? A recent analysis by King and Ebach (2017) sought to divide the biota within the Indo-Mayalan Transition Zone into separate Palaeogene and Neogene areas based on molecular clock age estimates. The result showed that regardless of geographical proximity, Palaeogene areas formed monophyletic clades separate from Neogene areas (Fig. 1.5). The formation of distinct temporal clades is evidence that the Indo-Mayalan Transition Zone is a geographical artefact due to two temporally disjunct overlapping areas. The areas west of Weber's line are Palaeogene in age, and the areas East of Wallace's line are Neogene in age, indicating that the Oriental region has a younger biota and the Australian region an older biota (Morrone and Ebach 2022).

1.4.4   Mexican Transition Zone (MTZ)

The Mexican Transition Zone includes the mountainous areas of Mexico, Guatemala, Honduras, El Salvador, and Nicaragua (Darlington 1957; Halffter 1964; Palestrini and Zunino 1986; Morrone 2015). The Mexican Transition Zone is subdivided by Morrone (2017) into five provinces: Sierra Madre Occidental, Sierra Madre Oriental, Transmexican Volcanic Belt, Sierra Madre del Sur, and Chiapas Highlands.

Morrone (2017) stated that because these provinces have characters either from Neartic or Neotropical regions, they belong simultaneously to both regions.

The Mexican Transition Zone was recognised as early as 1844–1846, by Johann Andreas Wagner in a three-part monograph published titled *Die Geographische Verbreitung der Säugethiere Dargestellt* (Wagner 1844, 1845, 1846), which was based on the area taxonomy of Illiger (1815). The series of papers includes the first global biogeographic map (see Wallaschek 2015; Egerton 2019), which is based on the animal distributions of Illiger (1815; Fig. 1.6). The map clearly shows the northern boundary of the Mexican Transition Zone, a fauna that Wagner (1844–1846) considered to be a mixture of Neartic and Neotropical taxa. Wagner noted the same observations made by Johann Karl Wilhelm Illiger and Heinrich Lichtenstein, who stated that 'South [of North America] is associated with the tropical South America' (Illiger 1815: 68) and 'It is the region of the earth in which the animal products of the north and south, the Alps and the tropical world, and indeed the eastern and western parts of the earth meet and intersect in the most wonderful way' (Lichtenstein 1830: 96).

Wallace also proposed the first regionalisation of the Neotropical region recognising four subregions: Mexican (southern Mexico and Central America), Antillean

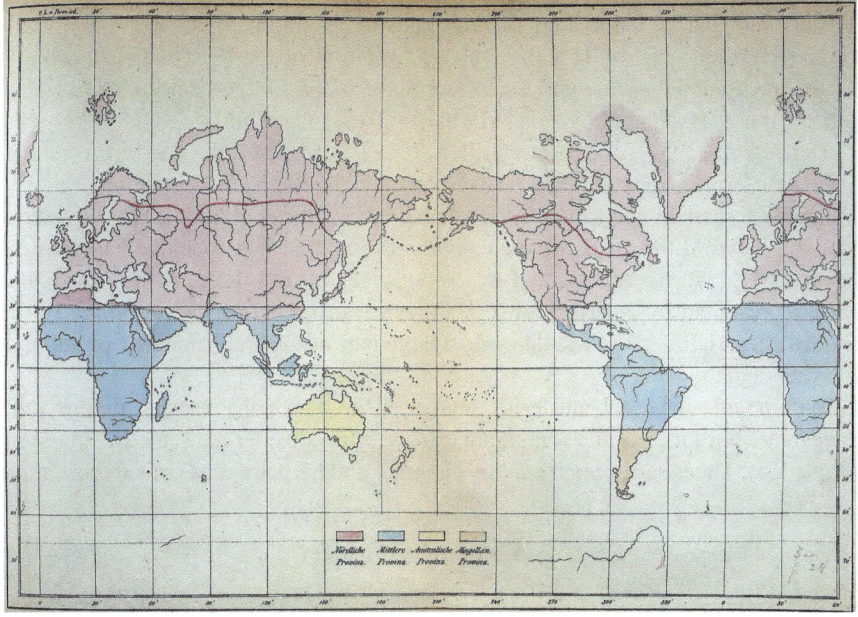

Fig. 1.6 The first known biogeographic map of the world, entitled 'Representation of the distribution of mammals according to their zones and their provinces. The southern boundary of the northern polar province is indicated by a line of a different colour, drawn somewhat further south than the equatorial border of the arctic fox (*Canis* [*Vulpes*] *lagopus*), though not so far in some places as the reindeer may descend there on their summer migrations. The southern polar province is not included in this map, because it is only in the process of discovery and, according to all previous experience, it does not harbour land mammals'. (Wagner 1846: 241)

(West Indies), Brazillian (tropical South America), and Chilean subregion (southern or temperate South America). Wallace (1876) considered Mexico transitional to the Nearctic, as well as Günther who considered the tropical fauna of southern Mexico advanced, 'is gradually mixed with the Artic fauna' (Günther 1858: 385). Allen (1892) noted, 'an interdigitation of the northern and southern life areas throughout the middle and southern portions of the North American continent' (Allen 1892: 205), which Merriam termed 'transition zones' (Merriam 1892: 31) and Lydekker (1896) called a subregion Wallace (1876). Mello-Leitao (1937) tweaked Wallace's classification and subdivided the Neotropical region into five subregions: Mexican (from Sonora in Mexico to Panama), Antillean (Antilles, excepted Trinidad and Tobago), Brazilian (South America west of Andes, from Colombia and Guyana to Bahia Blanca, Argentina), Andean-Patagonian (Andean Cordillera, from Colombia to Chile), and West Insular (Galapagos and Juan Fernandez Islands). Mello-Leitao (1937) also considered that the Mexican subregion has a mix of fauna coming from the Nearctic and Neotropical regions, similar to the phytogeographic transition zone of Vivo (1943).

Another contribution was made by Schmidt (1954) who proposed the Caribbean transition subregion, including the West Indian and Central American provinces, which he placed with the 'Holarctic rather than with the Neotropical Region. This action is based on their geologic history and on some faunal relations that are plainly Tertiary and perhaps not immediately evident' (Schmidt 1954: 327). Schmidt's claim is further supported by the Holarctic relation of the West Indian fossil and recent mammal fauna as well as the reptilian fauna of the 'Palaeopeninsula', a scheme that was adopted by Bond (1936) for West Indian birds.

Darlington (1957), and later Halffter (1962), recognised the Central American Transition Zone as containing a mix of Nearctic and the Neotropical biota, 'a zone of intense transition and superposition of faunas' (Halffter 1962: 2). He pointed out that there were differences between vertebrates and the Scarabaeidae fauna, as the beetles were mostly represented by Neotropical forms, whereas the vertebrates were represented by Neartic forms. Halftter's transition zone includes part of the southwestern USA, all of Mexico, and a large part of Central America extending to the Nicaraguan lowlands.

Within the analytical clustering analyses, the Mexican Transition Zone only appears within Proches and Ramdhani (2012), who consider it a separate subregion (Table 1.6). The cluster appears in the mammal and herpetofauna data and is almost

Table 1.6 Placement of the MTZ in existing area taxonomies

Authors	Realm/Kingdom	Region	Subregion
Schmidt (1954)	Arctogaean	Holarctic	Caribbean
Smith (1983)		Latin American	Neotropical
Cox (2001)	South American	–	–
Kreft and Jetz (2010)	–	Neotropical	Central America
Proches and Ramdhani (2012)	–	Neotropical	Central American
Holt et al. (2013)	Panamanian	–	–

absent in the bird data. The subregion also covers much of Central America, including areas not part of the Mexican Transition Zone, such as the eastern coastline of Mexico. Further investigation of the data would be necessary to see how the data clusters within the Central American region, given that the mammal data used by Kreft and Jetz (2010) found no Central American cluster; in fact, the entire tropical South and Central America fall into a single cluster, separate from the Antilles, Central Andes, and Patagonia. Holt et al. (2013) also failed to find any trace of the Mexican Transition Zones using amphibians, nonpelagic birds, and nonmarine mammals, in which Central America and the Caribbean fall into the Panamanian kingdom. Looking at the data in detail, namely, at taxon-specific analyses, Holt et al. (2013) showed that faunal turnover with the Mexican Transition Zone is highest in mammal and bird distributions. Based on these studies, the Mexican Transition Zone would potentially be an artefact of the overlap of two or more temporally disjunct areas.

1.4.5   South American Transition Zone (STZ)

The South American transition zone comprises the Andean highlands between western Venezuela and northern Chile and central western Argentina (Morrone 2004, 2006, 2014, 2015). It corresponds to the boundary between the Neotropical and Andean regions, which was analysed by Rapoport (1968), who discussed the alternative placements given by different authors to the 'subtropical line' that separates both regions (see also Ruggiero and Ezcurra 2003). There have been different definitions for the South American Transition Zone, such as the inclusion of the Patagonian Steppe (Roig-Juñent et al. 2018) and an entirely new area in the Pampean and Chaco provinces (Procheş and Ramdhani 2012). No other clustering approach has identified the South American Transition Zone (see Table 1.7), placing it in either the Neotropical or Andean province. Other than recent research (e.g. Morrone 2017, 2018), no other research discusses a South American Transition Zone.

Table 1.7 Placement of the STZ in existing area taxonomies

Authors	Realm/ kingdom	Region	Subregion
Schmidt (1954)	Neogaean	Neotropical	–
Smith (1983)		Latin American	Argentine
Cox (2001)	South American	–	Chilean
Kreft and Jetz (2010)	–	Neotropical	Patagonia, Central Andes, tropical S & C America
Procheş and Ramdhani (2012)	–	Andean/ Neotropical	–
Holt et al. (2013)	Neotropical	–	–

The South American Transition Zone may be an artefact from two geographically overlapping temporally disjunct regions as it does not form clusters in the similarity/dissimilarity analyses.

1.5   Transition Zones: Natural Area or Temporal Artefact?

Transition zones vary. In some cases, it is quite clear that the transition zone is artefactual, particularly in the case of the Indo-Malay TZ, which King and Ebach (2017) have shown to be two distinct temporal disjunct overlapping areas (Table 1.8). Even in the study by Procheş and Ramdhani (2012, table 2), the Indo-Malayan TZ ('Wallacea') only has three characteristic tetrapod genera, namely, sailfin lizards (*Hydrosaurus*), the parrot genus (*Tanygnathus*), and a genus of fruit bat (*Acerodon*). Of these, *Hydrosaurus* is also found in New Guinea (Australian region), and *Tanygnathus* is non-monophyletic and is considered a synonym of *Psittacula* (Podsiadlowski et al. 2017), which leaves *Acerodon* as the only true characteristic genus of tetrapods in Wallacea. Compared to the other regions of Procheş and Ramdhani (2012), the Wallacean region has the least support as a separate region, confirming the findings of King and Ebach (2017). It is clear that the CTZ and SATZ have by far the greatest support as natural areas, given that they form clusters within the analyses by Holt et al. (2013), Kreft and Jetz (2010), and Procheş and Ramdhani (2012), but further evidence is needed before they can be considered as separate natural areas within an area taxonomy.

A recent study by Ebach and Michaux (2020) incorporated a simple binary (e.g. absence/presence) matrix in which the tectonostratigraphic characteristics of transition zones were compared. The analysis recovered two groups: the Mexican, Indo-Malayan, and South American transition zones and the Saharo-Arabian and Chinese transition zones. These two groups are identical to the natural and artefact groupings in Table 1.8. Ebach and Michaux (2020) concluded that the result is possibly due to both groups of transition zones occurring along marginal and intra-plate boundaries, respectively. Clearly, more research on the role of neotectonics (i.e. dynamic topography and intra-plate stress fields) is needed to clarify the origins of biogeographical transition zones.

Morrone, like Major Forsyth before him, was careful to keep the transition zones out of area taxonomy. Their inclusion as hybrid areas only confuses matters, leading

Table 1.8 Transition zones categorised into potential natural areas or artefacts

Transition zone	Natural area	Artefact	Requires testing
Mexican		X	X
Saharo-Arabian	X		X
Chinese	X		X
Indo-Malayan		X	King and Ebach (2017)
South American		X	X

to conflicting taxonomies or aphyletic areas. Further study and development of a temporal approach used by King and Ebach (2017) are required to test these areas within a global comparative biogeographic analysis. Only when transition zones are resolved as separate areas will biogeographers be able to discover stable and monophyletic area taxonomies. The journey to finding a natural biogeographical classification is a vitally important goal for historical biogeography.

References

Allen JA (1892) The geographical distribution of north American mammals. Bull Am Mus Nat Hist 4:199–243

Arldt T (1907) Die Entwicklung der Kontinente und ihrer Lebewelt: ein Beitrag zur vergleichenden Erdgeschichte. W. Engelmann, Leipzig, p 729

Bond J (1936) Field guide to birds of the West Indies by James Bond. Academy of Natural Sciences of Philadelphia. Waverly Press, Baltimore

Boucot AJ, Johnson JG, Talent JA (1969) Early Devonian brachiopod zoogeography. Geol Soc Am Spec Pap 119:1–60. https://doi.org/10.1130/spe119-p1

Brenan JPM (1978) Some aspects of the phytogeography of tropical Africa. Ann Mo Bot Gard 65:437–478. https://doi.org/10.2307/2398859

Brummitt RK (2002) How to chop up a tree. Taxon 51:31–41

Burbidge N (1960) The phytogeography of the Australian region. Aust J Bot 8:75–211. https://doi.org/10.1071/bt9600075

Cox CB (2001) The biogeographic regions reconsidered. J Biogeogr 28:511–523. https://doi.org/10.1046/j.1365-2699.2001.00566.x

Crisci JV, Katinas L, Posadas P (2003) Historical biogeography: an introduction. Harvard University Press, Cambridge

Crisp MD, Cook LG, Steane DA (2004) Radiation of the Australian flora: what can comparisons of molecular phylogenies across multiple taxa tell us about the evolution of diversity in present-day communities? Philosophical transactions of the Royal Society of London. Ser B Biol Sci 359:1551–1571. https://doi.org/10.1098/rstb.2004.1528

Crisp MD, Trewick SA, Cook LG (2011) Hypothesis testing in biogeography. Trends Ecol Evol 26:66–72. https://doi.org/10.1016/j.tree.2010.11.005

Darlington PJ (1957) Zoogeography: the geographical distribution of animals. Wiley, New York, p 675

de Candolle A (1820) Essai élémentaire de géographie botanique. Dictionnaire des Sciences Naturelle Vol. 18. F. Levrault, Paris, pp 1–64

Deodati T, Cesarini D, Sbordoni V (2009) Molecular phylogeny, classification, and biogeographic origin of Callerebia and other related Sino-Himalayan genera (Insecta: Lepidoptera: Nymphalidae: Satyrinae). In: Hartmann M, Weipert J (eds) Biodiversität und Naturausstattung im Himalaya III (biodiversity and natural heritage of the Himalaya III). Verein der Freunde und Förderer des Naturkundemuseums Erfurt eV, Erfurt, pp 107–114

Di Virgilio G, Laffan SW, Ebach MC (2013) Quantifying high resolution transitional breaks in plant and mammal distributions at regional extent and their association with climate, topography and geology. PLoS One 8:e59227. https://doi.org/10.1371/journal.pone.0059227

Dickerson RE, Merril ED, McGregor RC, Schultze W (1928) Distribution of life in The Philippines, Monographs of the Bureau of Science, Manila, Philippine Island, Monograph, vol 21, pp 1–322

Dowding EM, Ebach MC (2016) The early Devonian palaeobiogeography of eastern Australasia. Palaeogeogr Palaeoclimatol Palaeoecol 444:39–47. https://doi.org/10.1016/j.palaeo.2015.11.037

Dowding EM, Ebach MC (2018) An interim global bioregionalisation of Devonian areas. Palaeobiodivers Palaeoenviron 98:527–547

Ebach MC (2015) Origins of biogeography—the role of biological classification in early plant and animal geography. Springer, New York. https://doi.org/10.1093/sysbio/syw024

Ebach MC (2017) Reinvention of Australasian biogeography: reform, revolt and rebellion. CSIRO Publishing, Melbourne. https://doi.org/10.1071/9781486304844

Ebach MC, Michaux B (2017) Establishing a framework for a natural area taxonomy. Acta Biotheor 61:127–135. https://doi.org/10.1007/s10441-017-9310-y

Ebach MC, Michaux B (2020) Biotectonics: tectonics as the driver of bioregionalisation. Springer, New York. https://doi.org/10.1007/978-3-030-51773-1

Ebach MC, Williams DM (2010) Aphyly: a systematic designation for a taxonomic problem. Evolut Biol 37:123–127

Ebach MC, Morrone JJ, Parenti LR, Viloria AL (2008) International code of area nomenclature. J Biogeogr 35:1153–1157. https://doi.org/10.1111/j.1365-2699.2008.01920.x

Ebach MC, Gill AC, Kwan AC, Ahyong ST, Murphy DJ, Cassis G (2013) Towards an Australian bioregionalization atlas: a provisional area taxonomy of Australia's biogeographical regions. Zootaxa 3619:315–342. https://doi.org/10.11646/zootaxa.3619.3.4

Ebmer AW (1998) Asiatische Halictidae—7. Neue Lasioglossum-Arten mit einer Übersicht der Lasioglossum s. str.-Arten der nepalischen und yunnanischenSubregion, sowie des nördlichen Zentral-China (Insecta: Hymenoptera: Apoidea: Halictidae: Halictinae). Linzer biologische Beiträge 30:365–430

Egerton FN (2019) History of ecological sciences, part 61B: terrestrial biogeography and paleo-biogeography, 1840s–1940s. Bull Ecol Soc Am 100:1–63. https://doi.org/10.1002/bes2.1465

Engler HGA (1899) Die Entwicklung der Pflanzengeographie inden letzen hundert Jahren. Wissenschaftliche Beiträge zum Gedächtniss der hundertjährigen Wiederkehr des Antritts von Alexander von Humboldt's Reisen ach Amerika. In: Anlass des Siebenten internat. Geographen Kongresses hrsg. Von der Gesellschaft für Erdkunde zu Berlin. W. H. Kühl, Berlin

Fabricius JC (1778) Philosophica Entomologica. Impensis Carol. Ernest. Bohnii, Hamburg

Ferro I (2024) Biogeographic transition zones. In: Scheiner SM (ed) Encyclopedia of biodiversity, vol 1, 3rd edn. Academic Press, Cambridge, MA, pp 460–465

Ferro I, Morrone JJ (2014) Biogeographic transition zones: a search for conceptual synthesis. Biol J Linn Soc 113:1–12. https://doi.org/10.1111/bij.12333

Forbes E (1846) On the connection between the distribution of the existing fauna and flora of the British Isles and the geological changes which have affected their area, especially during the epoch of the northern drift. Geol Surv Great Brit Mem 1:336–432

Forsyth Major CJ (1884) Zoogeographische Übergangsregionen. Kosmos 14:102–113

Fortelius M, Zhang Z (2006) An oasis in the desert? History of endemism and climate in the late Neogene of North China. Palaeontographica Abt A 277:131–141. https://doi.org/10.1127/pala/277/2006/131

Giraud Soulavie J-L (1780) Géographie de la nature, ou distribution naturelle des trois règnes sur la surface de la terre. Suivie de la Carte Minéralogique, Botanique, &c. du Vivarais où cette distribution naturelle est représentée. Ouvrage qui sert de préliminaire à l'Histoire Naturelle de la France Méridionale, &c. dont on va publier les deux premiers Volumes & à l'Histoire Ancienne & Physique du Globe Terrestre. Hôtel de Venise, Cloître Saint-Benoît, Paris. Et chez le Sieur Dupain-Triel, Ingénieur-Géographe du Roi, rue des Noyers

Good R (1953) The geography of the flowering plants. Longman Green & Co, London, p 403

Grisebach A (1866) Die Vegetations-Gebiete der Erde, übersichtlich zusammengestellt. Mittheilungen aus Justus Perthes' Geographischer Anstalt 12:44–53

Günther A (1858) On the geographical distribution of reptiles. Proc Zool Soc London 26:373–398

Halffter G (1962) Explicación preliminar de la distribución geográfica de los Scarabaeidae Mexicanos. Acta Zoológica Mexicana 5:1–17

Halffter G (1964) Las regiones Neártica y Neotropical, desde el punto de vista de su entomofauna, vol 1. Anais do II Congreso Latino-Americano de Zoología, São Paulo, pp 51–61

Heilprin A (1887) The geographical and geological distribution of animals. Appleton, New York

Holt BG, Lessard J, Borregaard MK, Fritz SA, Araújo MB, Dimitrov D, Fabre PH, Graham CM, Graves GR, Jønsson KA, Nogués-Bravo D, Wang Z, Whittaker RJ, Fjeldså J, Rahbek C (2013) An update of Wallace's zoogeographic regions of the world. Science 339:74–78. https://doi.org/10.1126/science.1228282

Huang C, Ebach MC, Ahyong ST (2020) Bioregionalisation of the freshwater zoogeographical areas of mainland China. Zootaxa 4742:271–298. https://doi.org/10.11646/zootaxa.4742.2.3

Humphries CJ, Ebach MC (2004) Biogeography on a dynamic earth. In: Lomolino M, Brown J (eds) Frontiers of biogeography: new directions in the geography of nature. Sinauer Press, Sunderland, pp 67–86

Illiger JKW (1815) Überblick der Säugthiere nach ihrer Vertheilung über die Welttheile. Abhandlungen der physikalische Klasse der Koeniglich-Preussischen Akademie der Wissenschaften 1804–1811:39–159

King AR, Ebach M (2017) A novel approach to time-slicing areas within biogeographic-area classifications: Wallacea as an example. Aust Syst Bot 30:495–512. https://doi.org/10.1071/sb17028

Kreft H, Jetz W (2010) A framework for delineating biogeographical regions based on species distributions. J Biogeogr 37:2029–2053. https://doi.org/10.1111/j.1365-2699.2010.02375.x

Kreft H, Jetz W (2013) Comment on "an update of Wallace's zoogeographic regions of the world". Science 341:343–343

Laffan SW, Lubarsky E, Rosauer DF (2010) Biodiverse, a tool for the spatial analysis of biological and related diversity. Ecography 33:643–647. https://doi.org/10.1111/j.1600-0587.2010.06237.x

Lichtenstein H (1830) Erläuterungen der Nachrichten des Franc. Hernandez von den vierfüssigen Thieren Neuspaniens. Druckerei der Königlichen Akademie der Wissenschaften 1830:89–127

Lydekker R (1896) A geographical history of mammals. Cambridge University Press, Cambridge

Mayr E (1944) Wallace's line in the light of recent zoogeographic studies. Q Rev Biol 19:1–14. https://doi.org/10.1086/394684

Mello-Leitao C (1937) Zoo-Geografia do Brasil. Companhia Editora Nacional, São Paulo

Merriam CH (1892) The geographical distribution of life in North America with special reference to the Mammalia. Proc Biol Soc Wash 7:1–64

Michaux B (2010) Biogeology of Wallacea: geotectonic models, areas of endemism, and natural biogeographical units. Biol J Linn Soc 101:193–212. https://doi.org/10.1111/j.1095-8312.2010.01473.x

Morrone JJ (2004) Panbiogeografía, components bióticos y zonas de transición. Rev Bras Entomol 48:149–162. https://doi.org/10.1590/s0085-56262004000200001

Morrone JJ (2006) Biogeographic areas and transition zones of Latin America and the Caribbean Islands based on panbiogeographic and cladistic analyses of the entomofauna. Annu Rev Entomol 51:467–494. https://doi.org/10.1146/annurev.ento.50.071803.130447

Morrone JJ (2014) Biogeographical regionalisation of the Neotropical region. Zootaxa 3782:1–110. https://doi.org/10.11646/zootaxa.3782.1.1

Morrone JJ (2015) Biogeographical regionalisation of the world: a reappraisal. Aust Syst Bot 28:81–90. https://doi.org/10.1071/sb14042

Morrone JJ (2017) Neotropical biogeography: regionalization and evolution. CRC Press, Boca Raton. https://doi.org/10.1201/b21824

Morrone JJ (2018) Evolutionary biogeography of the Andean region. CRC Press, Boca Raton. https://doi.org/10.1201/9780429486081

Morrone JJ, Ebach MC (2022) Toward a terrestrial biogeographical regionalisation of the world: historical notes, characterisation and area nomenclature. Aust Syst Bot 35:187–224

Müller S (1846) Bijdragen tot de kennis van Sumatra, bijzonder in geschiedkundig en ethnographisch opzigt. S. en J. Luchtmans, Leiden

Müller P (1972) Centres of dispersal and evolution in the neotropical region. Stud Neotropical Fauna Environ 7:173–185

Müller P (1979) Introducción a la Zoogeografía. Blume, Barcelona

Müller P (1986) Biogeography. Harper & Row, New York

Nelson G (1978) From Candolle to Croizat: comments on the history of biogeography. J Hist Biol 11:269–305. https://doi.org/10.1007/bf00389302

Nelson G, Platnick NI (1981) Systematics and biogeography: cladistics and Vicariance. Columbia University Press, New York

Olson DM, Dinerstein E (1998) The global 200: a representation approach to conserving the earth's most biologically valuable ecoregions. Conserv Biol 12:502–515. https://doi.org/10.1046/j.1523-1739.1998.012003502.x

Olson DM, Dinerstein E, Wikramanayake ED, Burgess ND et al (2001) Terrestrial ecoregions of the world: a new map of life on earth: a new global map of terrestrial ecoregions provides an innovative tool for conserving biodiversity. Bioscience 51:933. https://doi.org/10.1641/0006-3568(2001)051[0933:teotwa]2.0.co;2

Palestrini C, Zunino M (1986) L'analisi dell'entomofauna nelle zone di transizione: prospettive e problemi. Biogeographia 12:11–25. https://doi.org/10.21426/b612110278

Palestrini C, Simonis A, Zunino M (1985) Modelli di distribuzione dell'entomofauna della zona di transizione Cinese, analisi di esempi e ipotesi sulle sue origini. Biogeographia 11:195–209. https://doi.org/10.21426/b611110243

Palestrini C, Simonis A, Zunino M (1987) Modelli di distribuzione dell'entomofauna della Zona di Transizione Cinese, analisi di esempi e ipotesi sulle origini. Biogeographia 11:195–209

Parenti LR, Ebach MC (2009) Comparative biogeography: discovering and classifying biogeographical patterns of a dynamic earth. University of California Press, Berkeley/Los Angeles. https://doi.org/10.1525/9780520944398

Pelseneer P (1904) La Ligne de Weber, limite zoologique de l'Asie et de l'Australie. Bulletin de la classe des Sciences 9–10:1001–1022

Podsiadlowski L, Gamauf A, Töpfer T (2017) Revising the phylogenetic position of the extinct Mascarene parrot *Mascarinus mascarin* (Linnaeus 1771) (Aves: Psittaciformes: Psittacidae). Mol Phylogenet Evol 107:499–502. https://doi.org/10.1016/j.ympev.2016.12.022

Prichard JC (1826) Researches into the physical history of mankind, 2nd edn. Houlfton and Stoneman, London

Procheş S, Ramdhani S (2012) The world's zoogeographical regions confirmed by cross-taxon analyses. Bioscience 62:3. https://doi.org/10.1525/bio.2012.62.3.7

Rapoport EH (1968) Algunos problemas biogeográficos del Nuevo Mundo con especial referencia a la región Neotropical. In: Delamare Debouteville C, Rapoport EH (eds) Biologie de l'Amerique Australe, vol 4. Editions du Centre National de la Recherche Scientifique, Paris, pp 55–110

Rensch B (1930) Eine biologische Reise nach den Kleinen Sunda Inseln. Borntraeger, Berlin

Roig-Juñent SA, Griotti M, Domínguez MC, Agrain FA, Campos-Soldini P, Carrara R, Cheli G, Fernández-Campón F, Flores GE, Katinas L, Muzón JR, Neita-Moreno JC, Pessacq P, San Blas G, Scheibler EE, Crisci JV (2018) The Patagonian steppe biogeographic province: Andean region or south American transition zone? Zool Scr 47:623–629. https://doi.org/10.1111/zsc.12305

Rosen DE (1978) Vicariant patterns and historical explanation in biogeography. Syst Zool 27:159–188. https://doi.org/10.2307/2412970

Rosen BR (1988) From fossils to earth history: applied historical biogeography. In: Myers AA, Giller PA (eds) Analytical biogeography: an integrated approach to the study of animal and plant distributions. Chapman and Hall, London, pp 437–481. https://doi.org/10.1007/978-94-009-1199-4_17

Ruggiero A, Ezcurra C (2003) Regiones y transiciones biogeográficas: Complementariedad de los análisis en biogeografía histórica y ecológica. In: Morrone JJ, Llorente J (eds) Una perspectiva Latinoamericana de la biogeografía, Las Prensas de Ciencias. UNAM, Mexico, pp 141–154

Russel JR, Kniffen FB, Pruitt EL (1951) Culture worlds. The Macmillan Company/Collier-Macmillan Limited, London

Schmarda KL (1853) Die geographische Verbreitung der thiere. Carl Gerold and Son, Vienna

Schmidt KP (1954) Faunal realms, regions, and provinces. Q Rev Biol 29:322–331. https://doi.org/10.1086/400392

Schouw JF (1823) Grundzüge einer allgemeinen Pflanzengeographie. Reimer, Berlin

Simpson GG (1950) History of the Fauna of Latin America. Am Sci 38:361–389

Simpson GG (1977) Too many lines; the limits of the oriental and Australian zoogeographic regions. Proc Am Philos Soc 121:107–120

Smith CH (1983) A system of the world mammal faunal regions. I. Logical and statistical derivation of the regions. J Biogeogr 10:455–466. https://doi.org/10.2307/2844752

Smith BJ (1984) Regional endemism of the southeastern Australian land mollusc fauna. In: Solem A, van Bruggen AC (eds) World-wide snails: biogeographical studies on non-marine Mollusca. Backhuys, Leiden, pp 178–188. https://doi.org/10.1163/9789004631960_016

Stegmann B (1930) Die Vögel des dauro-mandschurischen Uebergangsgebietes. J Ornithol 78:389–471. https://doi.org/10.1007/bf01953148

Swainson W (1835) A treatise on the geography and classification of animals. Longman, Brown, Green, and Longmans, London

Takhtajan A (1986) Floristic regions of the world. University of California Press, Berkeley

Udvardy MDF (1975) A classification of the biogeographical provinces of the world. In: IUCN occasional paper (no. 18). International Union for Conservation of Nature and Natural Resources, Morges

Vivo JA (1943) Los Límites Biogeográficos en América y la Zona Cultural Mesoamericana. A Revista Geográfrica 7:109–131

von Humboldt A, Bonpland A (1807) Voyage de Humboldt et Bonpland (Première partie. Physique Générale, et relation historique du voyage. Premier Volume, Contenant Essai sur la Géographie des plantes, accompagné d'un Tableau physique des régions équinoxiales, et servant d'introduction à l'Ouvrage). Chez Fr. Schoell, Paris

Wagner A (1844) Die geographische Verbreitung der Saugethiere. Abhandlungen der Königlich Bayerischen Akademie der Wissenschaften. Math-Phys Kl 4:1–146

Wagner A (1845) Die geographische Verbreitung der Saugethiere. Abhandlungen der Königlich Bayerischen Akademie der Wissenschaften. Math-Phys Kl 4:37–108

Wagner A (1846) Die geographische Verbreitung der Saugethiere. Abhandlungen der Königlich Bayerischen Akademie der Wissenschaften. Math-Phys Kl 4:1–114

Wallace AR (1876) The geographical distribution of animals, vol I & II. Harper and Brothers, New York

Wallaschek M (2015) Johann Andreas Wagner (1797–1861) und die geographische Verbreitung der Säugthiere. Universitäts-und Landesbibliothek Sachsen-Anhalt

Wallace AR, Thiselton-Dyer WT (1885) The distribution of life. In: Humboldt library of popular science literature, vol 64. J. Fitzgerald & Co, New York, pp 1–247

Wickens GE (1976) Speculations on long distance dispersal and the flora of Jebel Marra, Sudan Republic. Kew Bulletin 31:105–150

Williams DM, Ebach MC (2020) Cladistics: a guide to biological classification. Cambridge University Press, Cambridge. https://doi.org/10.1017/9781139047678

Zimmermann EAW (1778–1783) Geographische geschichte des menschen, und der allgemein verbreiteten vierfüssigen thiere. Weygandschen Buchhandlung, Leipzig

Chapter 2
Impact Tectonics and Transition Zones: The Caribbean

Abstract We discuss the general characteristics of the Caribbean biota and identify and describe the different areas of endemism within the Caribbean. The tectonic history of the Caribbean Large Igneous Province (CLIP), which forms the core of the modern Caribbean Plate, is detailed from its genesis in the Late Cretaceous to the present day. We base our analysis on Pacific models that postulate its formation over the Galapagos hotspot in the Pacific and its subsequent translation eastwards to its present position. We use published molecular phylogenies to produce a general areagram to show the hierarchical relationship between Caribbean areas of endemism. This analysis identifies two clades with the Greater Antilles more closely related to the Bahama Banks/Florida, which in turn are sister areas to Central America and part of a North American clade, while the Lesser Antilles are more closely related to South America. We interpret this pattern in terms of biotectonic processes.

2.1 Introduction

Rodriguez-Silva and Schlupp (2021) recently reviewed the biogeography, past and present, of the Caribbean and highlighted important characteristics of its biota, such as its diversity, endemism, and the presence of many basal species. The Caribbean is usually interpreted as a biological transition zone between South and North American biotas. Ebach and Michaux (2020) classified transition zones into either intra-plate or marginal plate phenomena. Marginal plate transition zones can have a mixed biota because the boundary between continental plates in collision zones may not be sharp and well defined. In Wallacea, for example, a melee of rift fragments, marginal basins, and island arcs occupy the collision zones which have the potential to carry meta-populations from the source data.

Supplementary Information The online version contains supplementary material available at https://doi.org/10.1007/978-3-031-80162-4_2.

Our analysis of Caribbean biotectonics is structured around areas of endemism. Most islands have their own endemics, with large islands like Cuba and Hispaniola being important centres of endemism within the Caribbean. Island groups, such as the Greater Antilles (Cuba, Hispaniola, and Puerto Rico) and the twin arcs of the Lesser Antilles, are also endemic centres. Broadly speaking and bearing in mind the usual collection biases and complicating factor of being on an important migration route between the Americas, there are many more South American–related endemics than North American ones (Crews and Esposito 2020). Biodiverse groups are present in both plants, such as palms (135 species) and seed plants in general (13,000 species), and animals like capromyid rodents (32 species) and *Anolis* lizards (150 spp.). An extensive literature documents speciation and endemicity within the Caribbean (Rodriguez-Silva and Schlupp 2021), and it is neither our place nor our intention to attempt a synopsis. We provide a summary of the biota and geology for each island or island group.

The second section, based on Pacific models, describes how the Caribbean Large Igneous Province (CLIP) was formed to the west of its present position and then translated eastwards, reaching its present position at the end of the Paleogene. The space into which the CLIP crashed was a mix of proto-Caribbean rift terranes produced during the early phases of the Pangea breakup. We describe how Lower Cretaceous volcanism created an island arc system along CLIP's leading (eastern) edge and how proto-Atlantic rift terranes became accreted onto this margin. CLIP displaced the original ocean basin between the two cratons and eventually wedged itself along its northern and southern margins.

The third section is an analysis of the distributions of endemic taxa based on published molecular phylogenies. We describe how biological phylogenies are turned into areagrams, which are then combined into a general areagram by minimising conflict in the data. We discuss the implications of the hierarchical relationships displayed by the general areagram and present an account of the origin of the modern Caribbean Plate and its biota.

2.2   Areas of Endemism

2.2.1   *Bahamas*

The Bahamas include 29 inhabited islands and numerous smaller islands and cays which are the aerial parts of the Bahama Banks, a structurally complex feature of shallow carbonate banks separated by deeper channels. The Bahama islands are low-lying and composed of limestone and its erosion products, producing a landscape full of typical karst features including sinkholes and cave systems. The overall diversity of habitats is low because of the uniform substrate. Caribbean Pine Forest, covering almost a quarter of the land area, is renowned for its endemic birds and species shared with the Greater Antilles (Currie et al. 2019). Dry Broadleaf

Forest is also an important habitat and has its own endemic and restricted-range suite of species (Morrone 2014). The other habitat types include dry scrub, the dominant vegetation of the southern Bahamas, mangrove fringes, and wetlands. Wetlands are not common in karst landscapes but do occur on the larger islands and are rich in endemic taxa.

Hastings et al. (2014) described the modern vertebrate fauna as endemic at the species level with many taxa shared with the Greater Antilles. There is only one endemic bat, but the phylogenetic connection of the majority of the nine resident chiroptera is to the Greater Antilles (Speer et al. 2015); the terrestrial avifauna has shared endemic elements with the Greater Antilles (Arlott 2010) and Central America (Bond 1993); Knapp et al. (2011) listed three frogs (1), 25 lizards (13), 11 snakes (7), and two freshwater turtles present in the Bahamas (endemic numbers in brackets), which are part of a larger Greater Antillean fauna. Prior to human arrival, endemic tortoises and the terrestrial Cuban Crocodile were widespread, but the Bermudan species are now extinct, although tortoises and the crocodile can still be found in the Greater Antilles (Hastings et al. 2014).

The nature of the basement crust upon which the carbonate platforms have been built is unresolved because its intermediate thickness could indicate either a thinned continental crust or a thickened oceanic crust. Freeman-Lynde and Ryan (1987) and Masaferro and Eberli (1999) supported a Jurassic continental rift model with crustal thinning for the origin of the Bahama Banks. The modern carbonate platform has been built by corals since the start of the Paleogene and possibly for much longer and covers 17,000 km^2 and is up to 4500 m thick, a remarkable piece of environmental engineering by coral polyps and their algal guests. Morrone (2014) discussed two forest types—dry forest and pine forest—and listed Bahamian endemic species.

2.2.2 Jamaica

Jamaica, a medium-sized island (11,000 km^2), is mountainous in part, has a diverse array of different habitat types, and supports a distinctive biota. The island was once forested, but little of it now remains. Of the approximately 3000 native and naturalised plants, 827 are endemic with the flora showing strong phylogenetic links with Greater Antillean taxa (Kelly 1988). Habitat destruction and modification have greatly reduced diversity in many animal groups. Turvey et al. (2017) estimated that 29 mammal species have become extinct on the island since 1500 CE and that there may have been 100 endemic non-volant mammals in the Jamaican Holocene prior to human settlement. The surviving endemic non-volant mammalian fauna of Jamaica, the Greater Antilles, and the Bahamas consists of only 13 endemic species (one in Jamaica) placed in two small endemic families with distant South American origins. The herpetofauna (Stanely et al. 2021) and avifauna (Arlott 2010) are largely endemic and are diverse. Both are related to taxa endemic to the Greater Antilles plus the Bahamas, although Jamaican birds also show a strong Central

American signature (Bond 1993). The island is part of a great northern and southern movement of migratory birds, tracking food, travelling to breeding sites, or seeking warmer climes during the cold northern winters, and is an important component of the avifauna. Morrone (2014) provided additional descriptions of endemic areas and listed endemic species for each.

Jamaica is a Cretaceous island arc. The oldest rocks are Early Cretaceous in age, and fossil similarities link the arc to contemporary arcs now part of Central America (Mitchell 2021). If that is the case, then Jamaica is a detached part of the Central American Arc that makes up much of the Isthmus of Panama. But arcs of a similar age also occur in the Greater Antilles, in which case Jamaica could be a detached fragment of the Greater Antilles Arc. Later metamorphism, with upper-bound cooling ages ranging from the Late Cretaceous to Eocene, affected both the older volcanics and contemporaneous Palaeocene and Eocene clastic sediments. Coralline limestone platform building followed until uplift at the end of the Miocene (Abbott et al. 2013; Mitchell 2021). Today the island straddles a transform fault zone (Fig. 2.1). Morrone (2014, 2018) provided a list of endemic species and recognised two distinct endemic-rich environments: the dry forest of coastal and southern Jamaica and the cloud forests and other high-elevation habitats of central Jamaica.

2.2.3   The Greater Antilles

The Greater Antilles—Cuba, Hispaniola, and Puerto Rico (including the Virgin Islands)—are a group of larger islands. Their topography is diverse, from broad plains, swamps, and other wetlands to interior mountain ranges. The three islands were once thickly forested, but even today Caribbean pine forest, dry forest, rainforest, and mist or cloud forests on the higher peaks can be found throughout the Greater Antilles and Bermuda, and more distantly on the Central America Maya and Chortís terranes. The flora is both diverse and endemic. The different forest types have their own rich endemic fauna, some with South and Central American phylogenetic links. The islands straddle the great north-south migration route, the Atlantic Flyway, and the avifauna is numerically dominated by these seasonal visitors, mainly from the north, which account for 70% of the Cuban avifauna (Garrido and Kirkconnell 2000). Morrone (2014) recognised three centres of endemism in western, central, and southeast Cuba and listed their endemics.

2.2.4   Cuba

Cuba is the largest of the three main islands (110,000 km^2) and has a long geological history. The core of the island is a Cretaceous/Paleogene island arc system, but basement rocks exposed in west Cuba are of much older Grenville age (900 Ma) (Renne et al. 1989). They probably originated as rifted continental-margin terranes during the opening of the proto-Caribbean ocean between North and South America. It appears that the island arc was built on or near to such Proterozoic continental crust. Older Jurassic rocks occur along Cuba's northern margin and represent North American crust sutured to Cuba during its late Paleogene (45 Ma) collision with the North American Plate. The Neogene was dominated by carbonate platform building (Iturralde-Vincent et al. 2006). The Isla de la Juventud and several surrounding archipelagos are included with Cuba (Morrone 2018).

2.2.5   Hispaniola

The island of Hispaniola (Dominican Republic and Haiti) is a large island (76,000 km^2) with east-west trending mountain ranges that rise to over 3000 m and extensive plains. The diverse topography has led, in part, to a diversity of habitats, which even today support a rich endemic flora and fauna. Hispaniola's moist forest once covered more than half the island but is now sadly much reduced. Oak, mahogany, and palms are common components of lowland wet forests, and remnants still support a rich endemic fauna. Pine forest grows at mid-elevation in the mountain ranges and also supports a suite of local endemics. Dry forest makes up about 20% of the vegetation, principally in Haiti. It too supports a range of local endemics. Morrone (2014) listed Hispaniola's endemic species.

The island is a composite of a Late Cretaceous and younger arcs that collided with the Bahama Banks, the southern margin of the North American Plate, in the Paleogene (45 Ma) (Escuder-Virute et al. 2024). Neogene geology records long periods of carbonate platform building with renewed volcanic activity starting in the Pliocene. Kamenov et al. (2011) concluded that the lavas showed clear isotopic signatures of melts contaminated by deeply buried continental crust. These continental fragments are probably equivalents to those found in nearby Cuba.

2.2.6   Puerto Rico

Puerto Rico and the Virgin Islands (PR-VI Platform) from the eastern end of the Greater Antilles archipelago. Puerto Rico is the smallest of the Greater Antilles group (9000 km^2) but has a mountainous central region where elevations can exceed 1000 m, and broad coastal plains. Dinerstein et al. (1995) recognised two forest types: dry forest in the southern part of the island and moist forest in the central and northern parts. The dry forest is floristically distinct, with cacti being common, and is home to an endemic fauna. The moist forest is one of those rare examples where coverage has increased in modern times as farming declined and abandoned land has been left to regenerate. In the first fifty years following this change in economic direction, the area of cleared land allowed to revert to forest increased from just 6% to 41%, but continued logging of mature stands still occurs. Broad-leaf tropical rainforest is found at all elevations and has its own endemic fauna. Serpentinite areas in the west support a distinct flora that can tolerate the high heavy metal load of the soils. Morrone (2014) provided a list of Puerto Rico endemics.

Puerto Rico is an Early Cretaceous-Paleogene island arc complex overlain by Neogene limestone platforms. Older Jurassic continental-margin terranes are found in the central and northeast parts of the island (Larue 1994) and may represent North American crust transferred to the Caribbean Plate during their collision. In the west, serpentinites, cherts, and other units of a highly tectonised accretionary complex also date from the Jurassic (Montgomery et al. 1994). Modern Puerto Rico lies within a plate boundary zone (Fig. 2.1) between the Puerto Rico Trench to the north, where the Atlantic crust is subducting southwards, and in the south by the Muertos Trench, to the south where the Caribbean crust is subducting northwards (Mann et al. 2006). A zone of strike-slip faulting to the north of the island accommodates left-lateral movement.

2.2.7 Lesser Antilles

The Lesser Antilles are a group of small volcanic islands arranged as twin, parallel arcs. The outer arc islands are covered in limestone (Limestone Caribbees) and represent an extinct Late Cretaceous–Paleogene island arc complex (Wadge 1994).

Fig. 2.1 Major tectonic elements within the Caribbean region and the location and structure of the Caribbean Plate. *HPFZ* Hauncabamba-Palestina Fault Zone, *MSZ* Montague Suture Zone, *SCDB* South Caribbean Deformed Belt, *SSFZ* San Sebastian Fault Zone. Closed triangles represent subduction zones, open triangles = thrust zones. The Venezuelan and Columbian basins form the Caribbean Large Igneous Province (CLIP)

These islands still support a dry forest assemblage with an associated endemic fauna. The inner arc extends from Grenada in the south to Saba in the north and has been active since the Eocene. The islands support rainforests and, above 800 m, a cloud or elfin forest. The rainforest is floristically diverse and is home to an endemic fauna (Carrington et al. 2018). Acededo-Rodríguez and Strong (2008) concluded that the Lesser Antilles' flora showed a predominant South American influence. Morrone (2014) listed the endemics of the Lesser Antilles. The island arc complex has been built on ocean crust and becomes younger to the north, recording a complex and mostly localised volcanic history. A long period of arc activity has been generated by the westward subduction of Atlantic crust under the Caribbean Plate. There is some evidence that an earlier Cretaceous arc may underly much of the northern half of the Lesser Antilles, suggesting earlier phases of subduction around the eastern margins of the CLIP.

2.3 Tectonics of the Caribbean Plate

A locality map showing the major features of present-day tectonics is shown in Fig. 2.1. The Caribbean Plate is bounded by the North American, South American, Atlantic, and Cocos/Nazca plates and is largely composed of oceanic plateau crust (CLIP) with arcs, ophiolite belts, and microcontinental blocks attached to both its leading and lagging edges. The boundaries of the modern Caribbean Plate are clearly delimited in the east and west by subduction zones (the Lesser Antilles and Central American Arcs, respectively), but its boundaries have been obscured in the north and south by its interaction with the Americas. The Central American boundary is marked by a suture zone (Fig. 2.1: MSZ), which separates the Maya Block (North American Plate) from the Chortís Block (Caribbean Plate). The MSZ was part of the original boundary between the North American and Caribbean plates, which continued north to Cuba, Hispaniola, and Puerto Rico. The collision between the Greater Antilles and the Bahama Banks in the Paleocene (~ 45 Ma) resulted in Cuba suturing to the North American Plate. Oblique subduction continues further east beneath the northern margins of Hispaniola and Puerto Rico but has a strong shear component that now continues to the south of Cuba. The subduction zone connects to strike-slip fault zones in southern Hispaniola and Jamaica and spreads in the Cayman Trough. The southern boundary of the Caribbean and South American plates is defined by a broad deformed belt that includes parts of the northern coast of Venezuela, before it turns south along the eastern margin of the North Andes microplate.

Although in situ models for the origin of the Caribbean Plate have been suggested (James 2006, 2009, 2013; Keppie 2012, 2013), the consensus favours a Pacific origin and subsequent eastwards translation. Evidence supporting Pacific models comes from comparative biogeography (e.g. Escalante et al. 2007), palaeontology (e.g. Bandini et al. 2011), structural, stratigraphic, and magmatic histories (e.g. Kennan and Pindell 2009), and palaeomagnetic, seismic, and deep-sea drilling program data (e.g. Romito and Mann 2020). While the Caribbean Plate is usually interpreted as a single structure, Kerr and Tarney (2005) argued it was composite.

Figure 2.2 shows a series of cartoons tracking the development of the Caribbean Plate from the Late Cretaceous to the Neogene and is based on the reconstructions in Perelló et al. (2020). CLIP is a thickened oceanic plateau probably formed above the Galapagos hotspot (Fig. 2.2: GH) and subsequently carried eastwards as the Farallon Plate subducted under the Americas. The magma generated above the subduction zone created a Late Cretaceous volcanic arc that became attached to the eastern edge of CLIP during this translation. The earlier Jurassic opening of the proto-Caribbean (Fig. 2.2) had separated Gondwana from Laurasia. An initial rifting phase can produce detached continental fragments along the margins of the oceanic crust, and such terranes were also attached to CLIP's leading edge (Ross and Scotese 1988; Pindell et al. 2005).

The top right pane of Fig. 2.2 shows CLIP entering the active subduction zone, choking it and forcing a subduction polarity reversal as proto-Caribbean crust underthrusts CLIP. This also initiated subduction along CLIP's trailing edge, generating the intra-oceanic basalts of the Central American Arc (Pindell et al. 2005; García-Casco et al. 2008; Barbosa-Espitia et al. 2019). Continued eastward

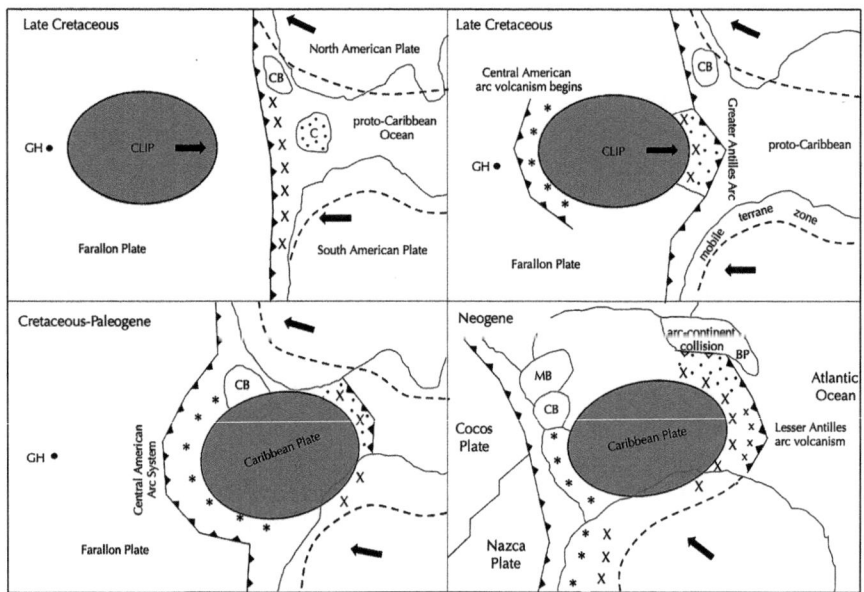

Fig. 2.2 Schematic illustration of the origin and development of the Caribbean Plate *C* Caribeana terrane, *CB* Chortís Block, *CLIP* Caribbean Large Igneous Province, *GH* Galapagos hotspot, *X* Great Caribbean Arc, *x* Lesser Antilles Arc, * Central American Arc, arrows indicate plate motion

movement of CLIP during the Cretaceous/Paleogene was accommodated by the subduction of the proto-Caribbean crust beneath it, until continued eastward movement was arrested by CLIP's interaction with both North and South America. The collision of CLIP's northern margin with North America was forceful enough to emplace ophiolites along parts of the Greater Antilles Arc, but the overall impression is that Caribbean Plate had lost much of its momentum by the end of the Paleocene and that the collision with the Bahama Banks was more like a docking. In Cuba and Hispaniola, the suture is marked by a zone of ophiolites and mélanges (Rojas-Agramonte et al. 2010; Escuder-Viruete et al. 2016). Protoliths of Palaeozoic continental crust, known from Cuba (Pindell et al. 2005), and Palaeozoic and Proterozoic zircons, from both Cuba (Renne et al. 1989; Rojas-Agromonte et al. 2010; Proenza et al. 2018) and Hispaniola (Torro et al. 2018), indicate that Grenville-aged (900 Ma) continental lithosphere is buried deep beneath much of the Greater Antilles. The collision shut down subduction along the Greater Antilles Arc and caused an eastward jump in active subduction to the inner Lesser Antilles Arc (Allen et al. 2019).

What part the Maya and Chortís blocks played in the collision of CLIP and the North American Plate is unclear. What is clear is that both are continental. The Maya Block is a composite (Weber et al. 2007); a possible Gondwanan terrane of Palaeozoic crust overlain by a Cretaceous limestone platform in the Yucatán, billion-year-old Grenville crust and Palaeozoic sediments representing a continental-margin terrane in the south, and the Chiapas Massif. The latter is a highly deformed and metamorphosed sediment pile originally derived from Grenville-aged source rocks and is another continental-margin terrane. The Chortís Block is also composite (Rogers et al. 2007)—a Neogene arc terrane, derived from the Central American Arc, is attached to a continental fragment of Grenville to Palaeozoic age bordered on the east by later Mesozoic basin terranes. Small, often elongated, continental fragments are products of rifting, and the Grenville-aged terranes are probably best treated as part of a rifted margin of one or both cratons along the margins of a newly opened proto-Caribbean Ocean. In the cartoons of Fig. 2.2, the Maya and Chortís blocks are conventionally placed close to North America, with the presence of other continental fragments southwards towards South America left purposefully vague. The boundary between the Caribbean and North American plates now passes between the Maya Block, sutured to the North American Plate, and the Chortís Block, which is part of the Caribbean Plate.

CLIP's southern border with the South American Plate is marked by a shear zone containing arc fragments and microcontinental blocks which stretch from northwest South America to the north Venezuelan coast (Boschman et al. 2014). The north Venezuelan coast is part of the southern margin of the Caribbean Plate (Neill et al. 2011; Escalona and Mann 2011; van der Lelij et al. 2010; Whattam and Stern 2015; Allen et al. 2019). Northwest South America is an amalgamation of Central American and Greater Antillean arc terranes sequentially accreted to the South American borderland during CLIP's eastwards passage, forming the North Andes Block (Cediel et al. 2003).

2.4 Data and Analysis

In this section, we describe the way species' distributions are combined with molecular phylogenies and analysed to search for hierarchical structure between areas. There are constraints limiting the usefulness, for our purposes, of published phylogenies. They must include geographically overlapping taxa that are not too widely distributed. Widespread taxa are not used because they are uninformative about area relationships and will only increase computational difficulties if not excluded. We restricted the 'in-area' to the Caribbean, northern South America, Central America (Panama, Costa Rica, Nicaragua, Guatemala), Mexico and southwest North America, and Florida. All distributions were compiled using Global Biodiversity Information Facility (GBIF) and additional online museum resources as needed. Species names were replaced by endemic areas in the phylogenies to produce individual areagrams, which were then combined to produce a general areagram showing what hierarchical structure was recovered.

The literature was surveyed for molecular trees of species distributed between two or more areas of endemism. Phylogenies were excluded if they contained distributions in the 'out area', if they were part of any basal, internal, or crown polytomy (Parenti and Ebach 2009) or included invasive or introduced species. Twenty-four molecular phylogenies matched these criteria, producing 39 clades involving 330 Caribbean species (Table 2.1; Electronic Supplementary Material). Phylogenies were converted into areagrams which were processed into paralogy-free subtrees (three-item statements) using the transparent method of Ebach et al. (2005). Compatibility analysis using the exhaustive branch and bound search option in LisBeth v 1.3 (Zaragüeta-Baglis et al. 2012) recovered the intersection tree, the general areagram, shown in Fig. 2.3 (RI = 0.65, Completeness Index = 84%).

2.5 General Areagram

2.5.1 Greater Antilles

The general areagram shown in Fig. 2.3 demonstrates that the Caribbean is non-monophyletic—the Lesser Antilles is a sister area to South America, while the Greater Antilles is part of a North American and Central American clade—but that it does resolve into two natural (monophyletic) groupings. The larger clade is composed of North America (north of the Yucatan Peninsula and southwest USA), Bahamas/South Florida, Greater Antilles, the Maya and Chortís blocks, and the Central American arc terrane.

The ingroup, linking the Greater Antilles and the Bahama Banks as sister areas, records the evolutionary consequences of the collision between the Caribbean Plate and the Bahama Banks and the range expansion of Caribbean species northwards. Whether there was a reverse process of North American/Bahama Banks species

Table 2.1 Molecular phylogenies used in the analysis of relationships between endemic areas within the Caribbean

References	Taxa
Andrus et al. (2009)	*Erigeron sp.*
Cacho and Baum (2012)	*Euphorbia sp.*
Cervantes et al. (2016)	*Tragia sp.; Adelia sp.*
Francisco-Ortega et al. (2007)	*Lasiocroton sp.*
Grose and Olmstead (2007)	*Tabebuia sp.*
Heinicke et al. (2007)	*Eleutherodactyline sp.; Craugastor sp.; Pristimantis sp.*
Lavin et al. (2001)	*Politea sp.*
Liu et al. (2004)	*Sachsia sp.*
McDowell et al. (2003)	*Exostema sp.*
McHugh et al. (2014)	*Micrathena sp.;*
Montelongo and Gómez-Zutita (2014)	*Caligrapha sp.; Zygospila sp.;*
Moynihan and Watson (2001)	*Neolaugeria sp.*
Negrón-Ortiz and Watson (2002)	*Erithalis sp.; Ernodea sp.*
Overton and Rhoads (2004)	*Turdo sp.*
Roncal et al. (2008)	*Coccothrinax sp.*
Silva et al. (2017)	*Amazona sp.*
Sturge et al. (2009)	*Icterus sp.*
Zhang et al. (2017)	*Apotomoderes sp.; Entiminae sp.; Lachnopus sp.; Diaprepes sp.; Exophthalmus sp.*

expanding across the collision zone is an interesting question. Cuba, which is now part of the North American Plate, would be a good place to search for that evidence. Any range expansion would have happened soon after a collision at ~45 Ma because the process takes place in ecological time, which is more-or-less instantaneous when viewed on a geological time scale. Ebach and Michaux (2020) argued that tectonically active plate boundaries promote speciation because the topography is dynamic and changes rapidly. Species are constantly being challenged by the frequency with which barriers to dispersal are made and broken; speciation on such landscapes is inevitable, as is a high extinction rate. The descendants of any Eocene taxa that expanded southwards should be local endemics related to more widespread northern clades.

The inclusion of both continental-rift terranes (Maya and Chortís blocks) and the Central American arc terrane in this clade supports Pacific models for the development of the Caribbean Plate. The Maya and Chortís blocks were part of a set of continental fragments formed around the margins of a proto-Caribbean Ocean during the early stages (Late Jurassic) of Laurasia's separation from Gondwana. The Central American Arc formed around CLIP's western margin as an intra-oceanic arc and now forms part of northwest South America and the Isthmus of Panama. A Neogene extension of these Central American arc volcanics occurred along the southern margin of the Chortís Block. When CLIP collided with these terranes, arc and Grenville-aged continental fragments became welded to its leading edge, which

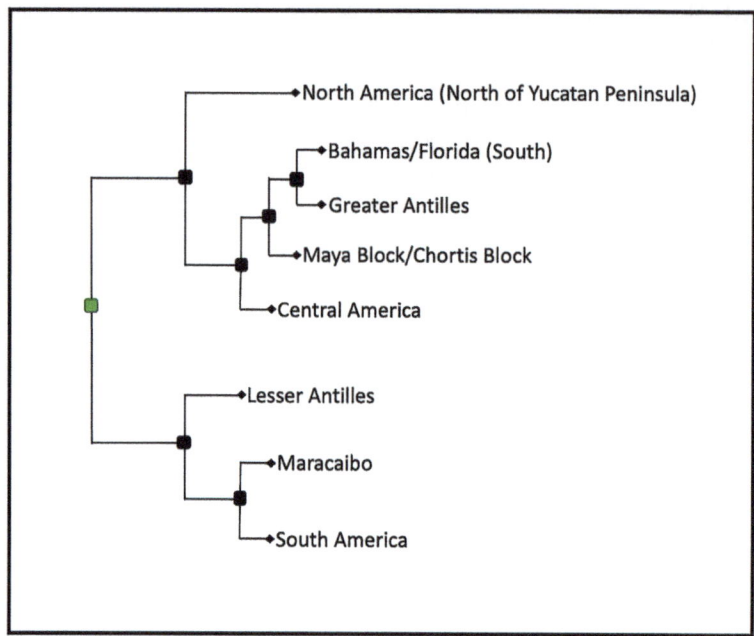

Fig. 2.3 General areagram of the Caribbean showing the relationship between the Caribbean provinces of Morrone (2018). North America = north of the Yucatan Peninsula, Maracaibo = north Venezuela coastal. Retention Index (RI) = 0.65, Completeness Index = 84%. RI represents the proportion of the three-item statements that support the intersection tree, and the Completeness Index is a measure of how many three-item statements implicit in the intersection tree are actually present in the original data set

we suggest is the origin of the strong biological connection between Central American (south of Mexico) and Caribbean life. We think most of the continental fragments were probably Gondwanan, although there is little geological evidence either way except for a possible Gondwana connection for parts of the Maya Block, but the Central American biota appears to be a markedly endemic offshoot of South America stock. A Gondwanan origin for the Maya and Chortís blocks does not preclude a Late Cretaceous position close to North America's southern margin because nothing is known of its Early Cretaceous tectonic history, and both could have been translated northwards along strike-slip faults.

2.5.2   Lesser Antilles

The Lesser Antilles are sister to a South American clade composed of Maracaibo and South America. Maracaibo is a strip of the north Venezuelan coast and offshore islands that stretches from northern Columbia to Trinidad and Tobago. The Caribbean mountains mark its southern extent. Maracaibo was a tectonically mobile zone that distributed the shear forces associated with CLIP's eastward movement along the northern margin of the Guiana Shield. The terranes of the mobile zone were separated from the thickened oceanic crust of CLIP by a subduction zone. Subduction was undoubtedly oblique, but parts of CLIP's southern margin must have been subducted beneath the terranes of the mobile zone, welding them to CLIP when subduction stopped ~ 45 Ma. The southern margin of the Caribbean Plate is now attached to South America.

The volcanics of Grenada, of much the same age, are a continuation of the Maraciabo mobile zone. The modern Lesser Antilles were formed above the westwards-subducting Atlantic crust. As this chain was built, South American and Marachiabo taxa would have expanded their ranges ever northwards along the arc of volcanic activity. The islands of the Lesser Antilles may not have all been built on oceanic crust because accreted Late Cretaceous arcs probably underly the outer arc Limestone Caribbees, much of the northern section of the Lesser Antilles and, to the west, the Aves Ridge. The present biota would have established itself along this active margin and expanded as opportunities arose from ~ 45 Ma.

2.6 Classifying the Caribbean Transition Zone

Ebach and Michaux (2020) classified transition zones into either intra-plate biotectonic (IPB) processes, associated with dynamic topography, or marginal plate biotectonic (MPB) processes. The Caribbean, Wallacean, and South American transition zones are all associated with plate margins and were grouped together in a MPB clade in the authors' cladistic analysis. An interesting feature of this grouping is the diversity of plate margin interactions involved: continent-continent (Wallacea), continent-ocean (STZ), and continent-CLIP-continent (Caribbean). The role played by the continents in Wallacea and the Caribbean differs; in Wallacea, they are in the early process of collision, while in the Caribbean they form the northern and southern boundaries of a trapped block of thickened oceanic crust (CLIP) that cannot be destroyed by subduction. Scattered terranes with Grenville-aged basements are now found in Central America and the Greater Antilles and would have originated as rift systems opened up and widened during the early stages of the opening of the proto-Caribbean Ocean. We cannot tell yet whether both cratons contributed rifted terranes, as is clearly the case in Wallacea, because the continental basement is too uniform to distinguish Laurasian from the Gondwanan crust. Biology, however, supports a southern origin for many of the endemics of the Caribbean Transition Zone. A second difference to Wallacea is that the impact on the zone of rift terranes/island arcs/small oceanic basins was provided by an allochthonous large igneous province (CLIP) rather than by continent-continent collision. The result of impact tectonics was that terranes of various sorts were transferred to CLIP's margins. The final stages in the development of the modern Caribbean Plate was its collision with the Bahama Banks, an amalgamation of its southern margin to South America, and the development of a westwards-dipping subduction along its eastern margin.

References

Abbott RN Jr, Bandy BR, Rajkumar A (2013) Cenozoic burial metamorphism in eastern Jamaica. Carib J Earth Sci 46:13–30

Acevedo-Rodríguez P, Strong MT (2008) Floristic richness and affinities in the West Indies. Bot Rev 74:5–36. https://doi.org/10.1007/s12229-008-9000-1

Allen RW, Collier JS, Stewart AG, Henstock T, Goes S, Rietbrock A et al (2019) The role of arc migration in the development of the Lesser Antilles: A new tectonic model for the Cenozoic evolution of the eastern Caribbean. Geology 47:891–895. https://doi.org/10.1130/g46708.1

Andrus N, Tye A, Nesom G, Bogler D, Lewis C, Noyes R, Jaramillo P, Francisco-Ortega J (2009) Phylogenetics of *Darwiniothamnus* (Asteraceae: Astereae)–molecular evidence for multiple origins in the endemic flora of the Galápagos Islands. J Biogeogr 36:1055–1069. https://doi.org/10.1111/j.1365-2699.2008.02064.x

Arlott N (2010) Birds of the West Indies. Collins, London

Bandini AN, Baumgartner PO, Flores K, Dumitrica P, Hochard C, Stampfli GM, Jackett S-J (2011) Aalenian to Cenomanian Radiolaria of the Bermeja complex (Puerto Rico) and Pacific origin of radiolarites on the Caribbean plate. Swiss J Geosci 104:367–408. https://doi.org/10.1007/s00015-011-0072-2

Barbosa-Espitia A, Kamenov GD, Foster DA, Restrepo-Moreno SA, Pardo-Trujillo A (2019) Contemporaneous Paleogene arc-magmatism within continental and accreted oceanic arc complexes in the northwestern Andes and Panama. Lithos 348-349:105185. https://doi.org/10.1016/j.lithos.2019.105185

Bond J (1993) Birds of the West Indies, 5th edn. Collins, London

Boschman LM, van Hinsbergen DJJ, Torsvik TH, Spakman W, Pindell JL (2014) Kinematic reconstruction of the Caribbean region since the early Jurassic. Earth Sci Rev 138:102–136. https://doi.org/10.1016/j.earscirev.2014.08.007

Cacho NI, Baum DA (2012) The Caribbean slipper spurge *Euphorbia tithymaloides*: the first example of a ring species in plants. Proc R Soc B Biol Sci 279:3377–3383. https://doi.org/10.1098/rspb.2012.0498

Carrington CM, Edwards RD, Krupnick GA (2018) Assessment of the distribution of seed plants endemic to the lesser Antilles in terms of habitat, elevation, and conservation status. Caribbean Nat Spec Issue 2:30–47

Cediel F, Shaw RP, Cáceres C (2003) Tectonic assembly of the Northern Andean block. In: Bartolini C, Buffler RT, Blickwede J (eds) The Circum-Gulf of Mexico and the Caribbean: hydrocarbon habitats, basin formation, and plate tectonics, vol 79. AAPG Memoir, pp 815–848. https://doi.org/10.1306/m79877c37

Cervantes A, Fuentes S, Gutiérrez J, Magallón S, Borsch T (2016) Successive arrivals since the Miocene shaped the diversity of the Caribbean Acalyphoideae (Euphorbiaceae). J Biogeogr 43:1773–1785. https://doi.org/10.1111/jbi.12790

Crews SC, Esposito LA (2020) Towards a synthesis of the Caribbean biogeography of terrestrial arthropods. BMC Evol Biol 20:12. https://doi.org/10.1186/s12862-019-1576-z

Currie D, Wunderle JM, Freid E, Ewert DN, Lodge DJ (2019) The natural history of The Bahamas: A field guide. Cornell University Press, New York. https://doi.org/10.1086/711804

Dinerstein E, Olson DM, Graham DJ, Webster AL, Primm SA, Bookbinder MP, Ledec G (eds) (1995) A conservation assessment of the terrestrial ecoregions of Latin America and the Caribbean. The World Bank, Washington, DC. https://doi.org/10.1596/0-8213-3295-3

Ebach MC, Michaux B (2020) Biotectonics: tectonics as the driver of bioregionalisation, Springer Briefs in Evolutionary Biology. https://doi.org/10.1007/978-3-030-51773-1

Ebach MC, Humphries CJ, Newman RA, Williams DM, Walsh SA (2005) Assumption 2: opaque to intuition? J Biogeogr 32:781–787. https://doi.org/10.1111/j.1365-2699.2005.01283.x

Escalante T, Rodríguez G, Cao N, Ebach MC, Morrone JJ (2007) Cladistic biogeographic analysis suggests a Caribbean diversification prior to the Great American Biotic Interchange and the Mexican Transition Zone. Naturwissenschaften 94:561–565

Escalona A, Mann P (2011) Tectonics, basin subsidence mechanisms, and paleogeography of the Caribbean-south American plate boundary zone. Mar Pet Geol 28:8–39. https://doi.org/10.1016/j.marpetgeo.2010.01.016

Escuder-Viruete J, Suárez-Rodríguez Á, Gabites J, Pérez-Estaún A (2016) The Imbert formation of northern Hispaniola: a tectono-sedimentary record of arc–continent collision and ophiolite emplacement in the northern Caribbean subduction–accretionary prism. Solid Earth 7:11–36. https://doi.org/10.5194/se-7-11-2016

Escuder-Viruete J, Fernández FJ, Pérez Valera F, McDermott F (2024) Active tectonics, quaternary stress regime evolution and seismotectonic faults in southern central Hispaniola: implications for the quantitative seismic hazard assessment. Geochem Geophys Geosyst 25:e2023GC011003. https://doi.org/10.1029/2023GC011003

Francisco-Ortega J, Santiago-Valentín E, Acevedo-Rodríguez P, Lewis C, Pipoly J, Meerow AW, Maunder M (2007) Seed plant genera endemic to the Caribbean Island biodiversity hotspot: a review and a molecular phylogenetic perspective. Bot Rev 73:183–234. doi:https://doi.org/10.1663/0006-8101(2007)73[183,spgett]2.0.co;2

Freeman-Lynde RP, Ryan WBF (1987) Subsidence history of the Bahama Escarpment and the nature of the crust underlying the Bahamas. Earth Planet Sci Lett 84:457–470

García-Casco A, Iturralde-Vinent M, Pindell J (2008) Latest cretaceous collision/accretion between the Caribbean plate and Caribeana: origin of metamorphic terranes in the greater Antilles. Int Geol Rev 50:781–809. https://doi.org/10.2747/0020-6814.50.9.781

Garrido OH, Kirkconnell A (2000) Birds of Cuba. Helm Field Guides, London. https://doi.org/10.1515/9781501751578

Grose SO, Olmstead RG (2007) Evolution of a charismatic neotropical clade: molecular phylogeny of *Tabebuia* sl, Crescentieae, and allied genera (Bignoniaceae). Syst Bot 32:650–659. https://doi.org/10.1600/036364407782250553

Hastings AK, Krigbaum J, Steadman DW, Albury NA (2014) Domination by reptiles in a terrestrial food web of The Bahamas prior to human occupation. J Herpetol 48:380–388. https://doi.org/10.1670/13-091r1

Heinicke MP, Duellman WE, Hedges SB (2007) Major Caribbean and central American frog faunas originated by ancient oceanic dispersal. Proc Natl Acad Sci 104:10092–10097. https://doi.org/10.1073/pnas.0611051104

Iturralde-Vinent MA, García-Casco A, Rojas-Agramonte Y, Proenza JA, Murphy JB, Stern RJ, James KH (2006) Arguments for and against the Pacific origin of the Caribbean plate: discussion, finding for an inter-American origin. In: Iturralde-Vinent MA, Lidiak EG (eds) Caribbean plate tectonics, vol 4. Geologica Acta, pp 279–302

James KH (2006) Arguments for and against the Pacific origin of the Caribbean Plate: discussion, finding for an inter-American origin. Geologica Acta 4:279–302

James KH (2009) In situ origin of the Caribbean: discussion of data. In: James KH, Lorente MA, Pindell JL (eds) The origin and evolution of the Caribbean plate, vol 328. Geological Society, Special Publications, London, pp 77–125. https://doi.org/10.1144/sp328.3

James KH (2013) Caribbean geology: extended and subsided continental crust sharing history with eastern North America, the Gulf of Mexico, the Yucatán Basin and northern South America. Geosci Can 40:3–8. https://doi.org/10.12789/geocanj.2013.40.001

Kamenov GD, Perfit MR, Lewis JF, Goss AR, Arevalo R Jr, Shuster RD (2011) Ancient lithospheric source for quaternary lavas in Hispaniola. Nat Geosci 4:554–557. https://doi.org/10.1038/ngeo1203

Kelly DL (1988) The threatened flowering plants of Jamaica. Biol Conserv 46:201–216

Kennan L, Pindell JL (2009) Dextral shear, terrane accretion and basin formation in the northern Andes: best explained by interaction with a Pacific-derived Caribbean plate? In: James KH, Lorente MA, Pindell JL (eds) The origin and evolution of the Caribbean plate, vol 328. Geological Society, Special Publications, London, pp 487–531. https://doi.org/10.1144/sp328.20

Keppie DF (2012) Derivation of the Chortis and Chiapas blocks from the western Gulf of Mexico in the latest cretaceous–Cenozoic: the pirate model. Int Geol Rev 54:1765–1775. https://doi.org/10.1080/00206814.2012.676356

Keppie DF (2013) The rationale and essential elements for the new 'pirate' model of Caribbean tectonics. Geosci Can 40:9–16. https://doi.org/10.12789/geocanj.2013.40.002

Kerr AC, Tarney J (2005) Tectonic evolution of the Caribbean and northwestern South America: the case for accretion of two late cretaceous oceanic plateaus. Geology 33:269–272. https://doi.org/10.1130/g21109.1

Knapp CR, Iverson JB, Buckner SD, Cant SV (2011) Conservation of amphibians and reptiles in the Bahamas. In: Conservation of Caribbean Island Herpetofaunas volume 2: regional accounts of the West Indies. Brill, Boston, pp 53–87

Larue DK (1994) Puerto Rico and the Virgin Islands. Caribbean geology, an introduction. In: Donovan SK, Jackson TA (eds) Caribbean geology; An Introduction. University of the West Indies Publisher's Association, Kingston, pp 151–165

Lavin M, Wojciechowski MF, Richman A, Rotella J, Sanderson MJ, Matos AB (2001) Identifying tertiary radiations of Fabaceae in the greater Antilles: alternatives to cladistic vicariance analysis. Int J Plant Sci 162:S53–S76. https://doi.org/10.1086/323474

Liu H, Trusty J, Oviedo R, Anderberg A, Francisco-Ortega J (2004) Molecular phylogenetics of the Caribbean genera *Rhodogeron* and *Sachsia* (Asteraceae). Int J Plant Sci 165:209–217. https://doi.org/10.1086/380746

Mann P, Rogers RD, Gahagan L (2006) Overview of plate tectonic history and its unresolved tectonic problems. In: Bundschuh J, Alvarado GE (eds) Central America: geology, resources and hazards, vol 1, pp 201–238. https://doi.org/10.1201/9780203947043.ch8

McDowell T, Volovsek M, Manos P (2003) Biogeography of *Exostema* (Rubiaceae) in the Caribbean region in light of molecular phylogenetic analyses. Syst Bot 28:431–441. https://doi.org/10.1043/0363-6445-28.2.431

McHugh A, Yablonsky C, Binford G, Agnarsson I (2014) Molecular phylogenetics of Caribbean Micrathena (Araneae: Araneidae) suggests multiple colonisation events and single Island endemism. Invertebr Syst 28:337–349. https://doi.org/10.1071/is13051

Mitchell SF (2021) Cretaceous geology and tectonic assembly of Jamaica. In: Davison I, Hull JNF, Pindell J (eds) The basins, Orogens and evolution of the Southern Gulf of Mexico and Northern Caribbean. Geological Society of London Special Publications 504. https://doi.org/10.1144/SP504-2019-210

Montelongo T, Gómez-Zurita J (2014) Multilocus molecular systematics and evolution in time and space of *Calligrapha* (Coleoptera: Chrysomelidae, Chrysomelinae). Zool Scr 43:605–628. https://doi.org/10.1111/zsc.12073

Montgomery H, Pessagno EA Jr, Lewis JF, Schellekens J (1994) Paleogeography of Jurassic fragments in the Caribbean. Tectonics 13:725–732. https://doi.org/10.1029/94tc00455

Morrone JJ (2014) Biogeographical regionalisation of the Neotropical region. Zootaxa 3782:1–110. https://doi.org/10.11646/zootaxa.3782.1.1

Morrone JJ (2018) Evolutionary biogeography of the Andean region. CRC Press, Boca Raton. https://doi.org/10.1201/9780429486081

Moynihan J, Watson LE (2001) Phylogeography, generic allies, and nomenclature of Caribbean endemic genus *Neolaugeria* (Rubiaceae) based on internal transcribed spacer sequences. Int J Plant Sci 162:393–401. https://doi.org/10.1086/319583

Negrón-Ortiz V, Watson LE (2002) Molecular phylogeny and biogeography of *Erithalis* (Rubiaceae), an endemic of the Caribbean Basin. Plant Syst Evol 234:71–83. https://doi.org/10.1007/s00606-002-0192-2

Neill I, Kerr AC, Hastie AR, Stanek K-P, Millar IL (2011) Origin of the Aves ridge and Dutch–Venezuelan Antilles: interaction of the Cretaceous 'Great Arc' and Caribbean–Colombian Oceanic Plateau? J Geol Soc Lond 168:333–347. https://doi.org/10.1144/0016-76492010-067

Overton LC, Rhoads DD (2004) Molecular phylogenetic relationships based on mitochondrial and nuclear gene sequences for the Todies (*Todus*, Todidae) of the Caribbean. Mol Phylogenet Evol 32:524–538. https://doi.org/10.1016/j.ympev.2004.01.004

Parenti L, Ebach M (2009) Comparative biogeography: discovering and classifying biogeographical patterns of a dynamic earth, vol 2. Univ of California Press. https://doi.org/10.1525/california/9780520259454.001.0001

Perelló J, García A, Creaser RA (2020) Porphyry-related high-sulfidation mineralization early in central American arc development: Cerro Quema deposit, Azuero peninsula, Panama. Bol Soc Geol Mex 72:a260719. https://doi.org/10.18268/BSGM2020v72n3a260719

Pindell J, Kennan L, Maresch WV, Stanek K-P, Draper G, Higgs R (2005) Plate-kinematics and crustal dynamics of circum-Caribbean arc-continent interactions: tectonic controls on basin development in proto-Caribbean margins. In: Avé Lallemant HG, Sisson VB (eds) Caribbean–South American plate interactions, Venezuela: geological society of America special paper, vol 394, pp 7–52. https://doi.org/10.1130/0-8137-2394-9.7

Proenza JA, González-Jiménez JM, Garcia-Casco A, Belousova E, Griffin WL, Talavera C, Rojas-Agramonte Y, Aiglsperger T, Navarro-Ciurana D, Pujol-Solà N, Gervilla F, O'Reilly SY, Jacob DE (2018) Cold plumes trigger contamination of oceanic mantle wedges with continental crust-derived sediments: evidence from chromitite zircon grains of eastern Cuban ophiolites. Geosci Front 9:1921–1936. https://doi.org/10.1016/j.gsf.2017.12.005

Renne PR, Mattinson JM, Hatten CW, Somin M, Onstott TC, Millán G, Linares E (1989) [40]Ar/[39]Ar and U-Pb evidence for late Proterozoic (Grenville) continental crust in north-Central Cuba and regional tectonic implications. Precambrian Res 42:325–341. https://doi.org/10.1016/0301-9268(89)90017-x

Rodriguez-Silva R, Schlupp I (2021) Biogeography of the West Indies: A complex scenario for species radiations in terrestrial and aquatic habitats. Ecol Evol 11:2416–2430. https://doi.org/10.1002/ece3.7236

Rogers RD, Mann P, Emmet PA (2007) Tectonic terranes of the Chortis block based on integration of regional aeromagnetic and geologic data. In: Mann P (ed) Geologic and tectonic development of the Caribbean plate boundary in northern Central America, vol 428. Geological Society of America, Boulder, pp 65–88. https://doi.org/10.1130/2007.2428(04)

Rojas-Agramonte Y, Kröner A, García-Cascoa A, Kemp T, Hegner E, Pérez M, Barth M, Liu D, Fonseca-Montero A (2010) Zircon ages, Sr-Nd-Hf isotopic compositions, and geochemistry of granitoids associated with the northern ophiolite mélange of Central Cuba: tectonic implications for late cretaceous magmatism in the northwestern Caribbean. Am J Sci 310:1453–1479. https://doi.org/10.2475/10.2010.09

Romito S, Mann P (2020) Tectonic terranes underlying the present-day Caribbean plate: their tectonic origin, sedimentary thickness, subsidence histories, and regional controls on hydrocarbon resources. In: Davison I, Hull JNF, Pindell J (eds) The basins, Orogens and evolution of the southern Gulf of Mexico and northern Caribbean. Geological Society, London. https://doi.org/10.1144/SP504

Roncal J, Zona S, Lewis CE (2008) Molecular phylogenetic studies of Caribbean palms (Arecaceae) and their relationships to biogeography and conservation. Bot Rev 74:78–102. https://doi.org/10.1007/s12229-008-9005-9

Ross MI, Scotese CR (1988) A hierarchical tectonic model of the Gulf of Mexico and Caribbean region. Tectonophysics 5:139–168. https://doi.org/10.1016/0040-1951(88)90263-6

Silva T, Guzmán A, Urantówka AD, Mackiewicz P (2017) A new parrot taxon from the Yucatán peninsula, Mexico—its position within genus *Amazona* based on morphology and molecular phylogeny. PeerJ 5:e3475. https://doi.org/10.7717/peerj.3475

Speer KA, Soto-Centeno JA, Albury NA, Quicksall Z, Marte MG, Reed DL (2015) Bats of The Bahamas: natural history and conservation. Bull Fla Mus Nat Hist 53:45–95. https://doi.org/10.58782/flmnh.afzn4036

Stanely L, Murray CM, Murray JJ, Crother BI (2021) Areas of endemism of Jamaica: inferences from Parsimony Analysis of Endemism based on amphibian and reptile distributions. Biogeographia 36:1–15

Sturge RJ, Jacobsen F, Rosensteel BB, Neale RJ, Omland KE (2009) Colonization of South America from Caribbean islands confirmed by molecular phylogeny with increased taxon sampling. Condor 111:575–579. https://doi.org/10.1525/cond.2009.080048

Torró L, Proenza JA, Rojas-Agramonte Y, Garcia-Casco A, Yang J-H, Yang Y-H (2018) Recycling in the subduction factory: Archaean to Permian zircons in the oceanic cretaceous Caribbean Island-arc (Hispaniola). Gondwana Res 54:23–37. https://doi.org/10.1016/j.gr.2017.09.010

Turvey ST, Crees JJ, Li Z, Bielby J, Yuan J (2017) Long-term archives reveal shifting extinction selectivity in China's postglacial mammal fauna. Proc R Soc B Biol Sci 284:20171979

van der Lelij R, Spikings RA, Kerr AC, Kounov A, Cosca M, Chew D, Villagomez D (2010) Thermochronology and tectonics of the leeward Antilles: evolution of the southern Caribbean plate boundary zone. Tectonics 29:TC6003. https://doi.org/10.1029/2009tc002654

Wadge G (1994) The Lesser Antilles. In: Donovan SK, Jackson TA (eds) Caribbean geology: an introduction. Univ West Indies, Jamaica, pp 167–178

Weber B, Iriondo A, Premo WR, Hecht L, Schaaf P (2007) New insights into the history and origin of the southern Maya block, SE México: U–Pb–SHRIMP zircon geochronology from metamorphic rocks of the Chiapas massif. Int J Earth Sci 96:253–269. https://doi.org/10.1007/s00531-006-0093-7

Whattam SA, Stern RJ (2015) Late cretaceous plume-induced subduction initiation along the southern margin of the Caribbean and NW South America: the first documented example with implications for the onset of plate tectonics. Gondwana Res 27:38–63. https://doi.org/10.1016/j.gr.2014.07.011

Zaragüeta-Baglis R, Ung V, Grand A, Vignes-Lebbe R, Cao N, Ducasse J (2012) LisBeth: new cladistics for phylogenetics and biogeography. Comptes Rendus Palevol 11:563–566. https://doi.org/10.1016/j.crpv.2012.07.002

Zhang G, Basharat U, Matzke N, Franz NM (2017) Model selection in statistical historical biogeography of Neotropical insects—the *Exophthalmus* genus complex (Curculionidae: Entiminae). Mol Phylogenet Evol 109:226–239. https://doi.org/10.1016/j.ympev.2016.12.039

Chapter 3
The South American Transition Zone

Abstract We identify and describe endemic areas within the Andes Range and adjacent areas that make up the South American Transition Zone (STZ), and analyse their relationships using published molecular phylogenies to produce a general are-agram. Two clades are well resolved, one linking the Páramo province (northern STZ) to Central America, which in turn is a sister group to the Caribbean forming part of a South American clade, and the other linking the Cuyan High Andes (southern STZ) to the Patagonia terrane. Both the northern STZ and southern STZ are transition zones. The central STZ is not well resolved, and we argue that it may be better interpreted as an ecogeographical endemic area until further evidence is presented. We interpret these three distinct parts of the STZ in terms of biotectonic processes. It is clear from our analysis that the STZ is non-monophyletic and composite.

3.1 Introduction

The South American Transition Zone (STZ) lies along the western margin of the continent (Fig. 3.1). It comprises the Andean Range from western Venezuela to northern Chile, the desert areas of coastal Peru, and central western Argentina (Morrone 2006, 2014, 2015, 2017, 2018). Habitat diversity is high with topography rising from sea level to mountains almost 7000 m high, and prodives habitats to support a diverse and largely endemic biota. STZ is divided into seven endemic areas:

- Páramo: the high cordilleras of Venezuela, Colombia, Ecuador, and Peru >3000 m elevation (Fig. 3.1:1).
- Desert: a narrow strip along the Pacific Ocean coast, from northern Peru to northern Chile (Fig. 3.1:2).

Supplementary Information The online version contains supplementary material available at https://doi.org/10.1007/978-3-031-80162-4_3.

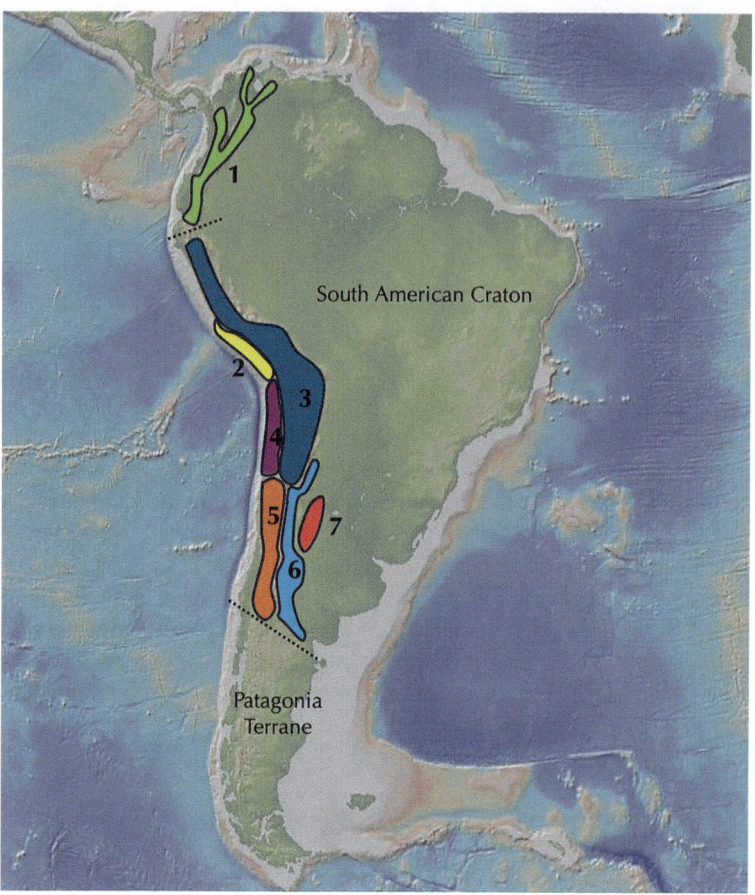

Fig. 3.1 Areas of endemism within the South American Transition Zone (STZ). Modified from Morrone (2018). 1 = Páramo province, 2 = Desert province, 3 = Puna province, 4 = Atacama province, 5 = Cuyan High Andes, 6 = Monte province, 7 = Comechingones province. The Patagonia terrane is a Palaeozoic microcontinent of unsettled affinity and amalgamation age

- Puna: the cordilleras and high plateaux of southern Peru, Bolivia, northern Argentina, and northern Chile (Fig. 3.1:3).
- Atacama: the coastal desert of northern Chile, between 18°S and 28°S (Fig. 3.1:4).
- Cuyan: the high Andes of Argentina and central Chile (30°–40°S) >2200 m elevation (Cabrera 1971; Wardle et al. 2001) (Fig. 3.1:5).
- Monte: the eastern foothills and plains of southern Bolivia and central Argentina (Fig. 3.1:6).
- Comechingones: isolated mountains of central Argentina (29°S–33°S) > 1000 m elevation (Fig. 3.1:7).

Ramos (2009) divided the Andean orogenic cycle into seven parts:

- Quiescence with no deformation and incipient arc magmatism.
- Growth of the magmatic arc and increasing compression and deformation: Based on the works of Kay et al. (1987), Kay et al. (1988), and Mpodozis and Ramos (1990), Ramos (2009) established that in some places, magmatic activity had extended towards the foreland, while, in other places, magmatic activity was restricted to the arc. Some areas may have undergone crustal erosion because of subduction, while other areas remained stable.
- Shallowing of the subducting slab and increased crustal thickening in the over-riding slab: The areas where shallowing occurred were associated with arc extension between 30°S and 33°S, where thermal weakening promoted foreland fragmentation.
- Flat-slab subduction, magmatic lull, and deformation: Areas where there was flat-slab subduction could lead to a coupling of upper and the lower plates resulting in intra-crustal deformation, although shallowing subduction did not always lead to a flat-slab phase (Ramos and Folguera 2005).
- Steepening of subduction and extensional collapse: in some areas, such as the southern part of Mendoza (35°S–36°S), slab steepening is associated with extensive basalt magmatism in the foreland as well as significant extensional collapse. Areas underlain by thick crust, such as the Altiplano or Puna, were largely unaffected (Gotze and Krause 2002; Beck and Zandt 2002).
- Retreat of arc magmatism and crustal and lithospheric delamination: The retreat of a magmatic arc on thickened crust is often accompanied by the development of calderas and ignimbrite flows (De Silva 1989; Kay et al. 1999; Caffe et al. 2002; Zandt et al. 2003), and this widespread dacitic magmatism is associated with plateau uplift (Coira et al. 1993; Gubbels et al. 1993; Beck and Zandt 2002; Garzione et al. 2006). The dacite melts are associated with crustal delamination.
- Uplift and final foreland deformation.

3.2   Areas of Endemism within STZ

3.2.1   *Páramo province*

The high cordilleras of Venezuela, Colombia, and Ecuador (3000–5000 m) define this province, which is known for its distinctive flora and highly biodiverse fauna. The tropical montane forest supports high numbers of endemics and gives way altitudinally to upland moorland, known as Páramo, which grows between the tree and snow lines. It too supports a distinctive, diverse, and endemic biota (Hughes and Eastwood 2006; Morrone 2014). A lower altitude shrub/grass flora grows in drier habitats. Four altitudinal zones have been recognised:

- High Andean mid-montane forests between 3000m and 3200 m.
- Shrub and scrub zone in drier lower altitudes 3200m and 3500 m.

- Páramo between 3500m and 4100 m. High biodiversity with a strong presence of xeromorphic vegetation. Grassland species, like *Calamagrostis, Festuca*, and dwarf bamboos (*Chusquea*) dominate the wetter slopes. It is also rich in shrubs, rosettes (*Acaena*), and cushion plants (*Werneria*) (Morrone 2018).
- Super-Páramo above 4100 m. Reaches the lower limit of the snow line and is characterised by patchy vegetation and bare ground. The most common plants are species of *Draba, Ephedra, Lupinus*, and *Senecio* (Morrone 2018).

3.2.2   Desert province

The Desert province is a narrow (20–30 km), arid strip along the Pacific coast of Peru. The coastal plain, while generally of low relief, does contain ridges and other elevated areas <600 m high. The Peruvian Desert grades into the Atacama Desert of Chile further south forming a 3500-km-long arid zone. The Desert province supports a surprisingly diverse flora because although it is one of the driest places on earth, winter fog provides enough moisture to support plant communities (lomas) on elevated ridges where moisture is trapped (Rundel et al. 1991). Rundel et al. (1991) described these lomas communities, noting their high levels of endemism (overall about 40% but as high as 60% in southern Peru) and species richness (~557 spp.). The flora consists of small trees and shrubs where sufficient moisture is present but can be characterised as herbaceous, salt tolerant, and xerophytic. Cacti are prominent. Although the Peruvian and Atacama deserts are contiguous, there is a sharp floristic disjunction between them. The diversity of plants is noticeably lower in the Atacama Desert (~373 spp.) and only 7% of species occur on either side of the disjunction. Morrone (2014) provided a list of endemic animals.

3.2.3   Puna province

The Puna province covers the High Andean Plateau, the Altiplano (3800–4500 m) of southern Peru and northwestern Argentina (15 °S–27 °S), and is home to a variety of mammals, such as an endemic camelid, the Vicugna, and a diverse endemic avifauna (Ramirez et al. 2007). The climate is cold and frosty (mean temperature < 9°C) with seasonal extremes, dry, and has high UV levels (Rojo et al. 2019). Despite these harsh conditions, two plant communities have established wherever moisture levels allow: a shrub-steppe composed of herbaceous, occasionally woody, plants and tussock grasses and a low-growing xerophytic assemblage including extensively developed fescue grasslands. Arzamendia and Vilá (2015) provided species lists for such a site in Argentina. These plant communities support a diverse and endemic fauna and are classed as a biodiversity hotspot. Morrone (2014) provided a list of endemic species.

3.2.4   *Atacama province*

The Atacama province of northern Chile (18°S–26 °S) is in many ways a continuation of the Peruvian deserts to the north. Physically, they are very similar with broad flat plains broken in places by higher-elevation land. These high spots intercept the winter fogs that sweep in from the Pacific allowing a lomas flora to develop. Away from these fog zones, the flora becomes less diverse, and the lomas gives way to a community of succulent cacti and low, semi-deciduous shrubs under taller, drought-tolerant shrubs like *Euphorbia*. There is also a flora restricted to wetter areas that can form local hotspots of diversity. But there are differences too—the Atacama province has a lower plant diversity and shares few species with the Peruvian deserts. Morrone (2014) provided a list of the province's endemics.

3.2.5   *Cuyan High Andean province*

The Cuyan High province includes the eastern slopes of the Argentinian Andes and mountains of central Chile (27°S–38°S). The Andes reach their highest elevation in the north (Mt. Aconcagua at 6960 m) decreasing to ~3000 m in the south (Ezcurra et al. 2020). The High Andes area of endemism is found above the tree line and is a cold, dry, and exposed environment. The flora is surprisingly diverse considering the harshness of the environment, with 381 species of vascular plants reported by Padro et al. (Padró et al. 2020). Perennial grasses dominate with shrubs and associated plants, but dwarf trees and cushion plants are also common. Similar communities with close relatives live to the south in Patagonia, as do freshwater fish (López et al. 2008) and a variety of invertebrate taxa (Roig-Juñent et al. 2018). Morrone (2006) did not classify the Cuyan High Andes as a part of STZ and regarded it as part of Patagonia. Morrone and Ezcurra (2016) provided a list of endemic species.

3.2.6   *Monte province*

The Monte province is located to the east of the Andes (24°S–44°S) and stretches some 2000 km from southern Bolivia to central Argentina (Morrone 2006, 2014; Roig et al. 2009; Elías and Aagesen 2016). It is an arid region that extends from sea level to 3500 m. Roig et al. (2009) recognised a number of areas of endemism based on insects, reptiles, mammals, and birds and described three vegetational types. A narrow strip of true desert (50–100 mm rainfall) is found in the high valleys of the Andean foothills and has vegetation dominated by evergreen shrubs, cacti, and annuals. In the north, where there is higher rainfall (200–400 mm) and a closed canopy forest is developed along water courses, which is replaced by shrubland and perennial herbs in drier areas. The southern area is also wetter (100–500 mm) and is characterised by creosote shrublands. Levels of endemism are high with numerous endemic genera and tribes of plants. Roig et al. (2009) quoted average figures of 30% endemism for insects and reptiles, 21% for mammals, and 12% for birds.

While there are a few plant genera that are also found in the climatically and geomorphologically similar Sonora and Mojave deserts to the north in Mexico and the southwestern USA, the majority of the flora and fauna are a mixture of Patagonian and Neotropical species, with a smaller palaeoendemic Austral element whose closest relatives are on other southern lands. Morrone 2014 provided a list of endemic species.

3.2.7   Comechingones province

The Comechingones province corresponds to an isolated mountain range to the east of the Andes in central Argentina (Martínez et al. 2017). The summers are short and cool, often with high humidity, and the winters are long, cold, and dry. The mean annual temperature at 1500 m elevation is 10°C, and the annual rainfall is 890–1400 mm (Goleniowski et al. 2006). Goleniowski et al. (2006) recognised three vegetation zones. On sunny slopes <1800 m, the predominant vegetation is *Schinopsis* forest, which often grows in dense stands and shrubland. On the cooler, on shady slopes <1800 m, the *Schinopsis* forest is replaced by *Lithraea* forest, and above 1800 m, the forest gives way to grasslands. Martínez et al. (2017) provided a list of endemic species.

3.3   Tectonics of the South American Transition Zone

The South American Transition Zone (STZ), shown in Fig. 3.1, occupies a large part of the Andean Ranges extending for more than 8000 km along the Pacific margin of South America from the Caribbean Sea to Tierra del Fuego. Gansser (1973) divided the Andes mountains into three parts: Northern, Central, and Southern Andes.

3.3.1   Northern Andes

Figure 3.1:1 shows the Northern Andes, which consists of an arc–arc–continental margin–craton sequence (Spikings et al. 2015). Its complex geological structure is detailed in Fig. 3.2. Grenville-aged continental inliers within a shear zone represent continental margin terranes within a collision mélange of Triassic–Jurassic arc material. Two younger arcs were accreted to this margin when CLIP collided with the northwest margin of South America. Greater Antilles Arc material was accreted

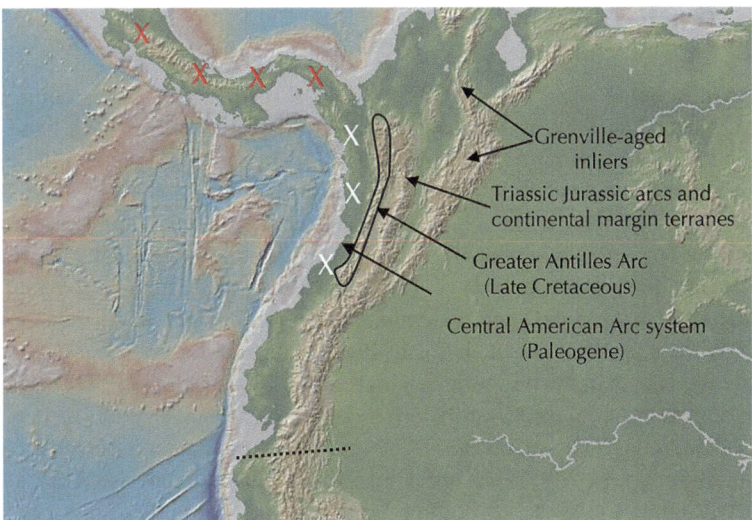

Fig. 3.2 Locality map of northwest South America showing the Páramo province in detail. White cross = Central American Arc System (Paleogene, extinct), red cross = Central American Arc System (Neogene, recent)

onto the South American Plate during the initial stages of the collision (Fig. 2.2), and similar-aged Late Cretaceous arc material transferred from its lagging edge (Central American Arc) towards the end of the collision. This arc is still active to the north in Central America.

3.3.2 Central Andes

The Central Andes (Fig. 3.1:3) extends from the Peruvian-Ecuadorian border for some 5200 km. It reaches elevations in excess of 6000 m and is built on a crust <70 km thick (Allmendinger et al. 1997; Bohm et al. 2002; Isacks 1988). It is divided into the Western and Eastern Cordilleras separated by a high plateau known as the Altiplano or Puna. The Western Cordillera records a long period of compression and subduction-related arc magmatism of Late Cretaceous to Paleogene age (Quade et al. 2015). According to Quade et al. (2015), it had reached elevations of 3–4000 m by the Paleocene. Neogene volcanic activity in the Central Andes built impressive stratovolcanoes and produced extensive lava flows and ignimbrites eruptions. The Eastern Cordilleras are non-volcanic and are composed of old Proterozoic and Palaeozoic continental basement. Uplift in the Neogene raised it to heights >3500 m as the deformation front spread eastwards. Between the two cordilleras lies the Altiplano or Puna, a high plateau region (~ 4000 m), formed by mid-Paleogene to mid-Neogene uplift of Palaeozoic continental basement. Extensive Neogene volcanics are also present on the plateau.

3.3.3   Cuyan High Andes

The Cuyan province lies between 30°S and 40°S (Fig. 3.1: 5) and is home to the Andes' tallest peak, Mount Aconcagua (6961 m) and is part of a long-lived compression zone dating from the mid-Mesozoic (Giambiagi et al. 2011). It is divided into two parts: a magmatically quiescent section north of 32°S that reaches elevations >6000 m, the highest mountains in the world not formed by continent-continent collision, and a southern section of lower elevation that is magmatically active. Like the Altiplano, the unusual heights of the northern section are thought to result from Neogene deformation and rapid exhumation due to flat-slab subduction, possibly caused by the subduction of the buoyant Juan-Fernandez Ridge (Pons et al. 2023). Crustal thicknesses in the northern section reaches 70 km.

Arc activity extends southwards along the western boundary of the Patagonia terrane. The Andes here are lower (< 3000 m) and have a thinner basement (average 35 km) (Bohm et al. 2002) than the High Andes. Subduction-related arc activity was, according to Cembrano and Lara (2009), stationary from the Jurassic to late in the Paleogene, when the modern arc activity migrated eastwards present active phase started. The earlier arc system was built on a continental basement composed of a Proterozoic core—the North Patagonian and del Deseado Massifs representing continental fragments formed during the early stages of Gondwana's breakup—and Paleaozoic metamorphics and plutons with thick Mesozoic sedimentary sequences. Late Cretaceous subduction produced little magmatic activity along this continental margin due to low angle or flat-slab subduction (Ramos et al. 2014), until 'normal' subduction resumed late in the Paleogene with the modern volcanic arc dating from ~30 Ma (Kay et al. 2013).

3.4   Biotectonics of the South American Transition Zone

3.4.1   Material and Methods

The areas used in this study are based on the bioregionalisation of the Neotropical region of Morrone (2014, 2017, 2018). The STZ provinces investigated include Paramo, Desert, Puna, Atacama, Cuyan High Andean, and Monte. The small Comechingones province was excluded from the analysis because of the lack of data. All species distributions were compiled using the Global Biodiversity Information Facility (GBIF) as well as other online museum databases as necessary.

The literature was surveyed for molecular trees that covered two or more biogeographical provinces of the STZ. The following criteria were used to reduce area paralogy and MASTs: (i) All non-endemic occurrences (i.e. widespread occurrences outside the region or subregion of a taxon known distribution) were excluded, (ii) single occurrences of widespread distributions were excluded (see Murphy et al. 2019), (iii) all polytomies (basal, internal, and crown) were excluded (see Parenti

Table 3.1 Molecular phylogenies used in the analysis of relationships between endemic areas within the South American Transition Zone

References	Taxa
Bell et al. (2012)	*Valeriana sp*
Ceccarelli et al. (2016)	*Brachistosternus sp*
Ferreti et al. (2012)	*Hapalopus sp.; Cyriococmus sp*
Jara-Arancio et al. (2017)	*Leucheria sp*
Meerow et al. (2020)	*Hymenocallis sp*
Gutiérrez Morales et al. (2020)	*Bletia sp.; Cattleya sp.; Epidendrum sp.; Neocogniauxia sp.; Dilomilis sp.; Brachionidium sp.; Octomeria sp.; Sansonia sp.; Pleurothallopsis sp.; Dresslerella sp.* and *Myoxanthus sp*

and Ebach 2009), (iv) familial level phylogenies were excluded (i.e. in which the smallest taxa are genera), and (v) known commercial crops or invasive or introduced species were excluded.

A literature search found 254 taxa from 27 molecular trees that matched these criteria (Table 3.1; Electronic Supplementary Material). These molecular trees were used in a paralogy-free subtree analysis to extract subtrees. The subtrees were converted into a three-item matrix and combined using compatibility analysis to find a single minimal tree using the program LisBeth (Zaragüeta-Bagils et al. 2012).

3.4.2 Results

The analysis found 256 subtrees and 248 compatible minimal trees from 22 of 27 molecular trees (Electronic Supplementary Material). Five of the original 27 trees (*Valeriana*, *Leucheria*, and *Hymenocallis*) contained unresolvable MASTs and were removed. Compatibility analysis using the exhaustive branch and bound search option in LisBeth v 1.3 (Zaragüeta-Bagils et al. 2012) recovered the intersection tree (general areagram) shown in Fig. 3.3 (RI = 0.62, Completeness Index = 24%).

The general areagram reveals two clades: the larger clade identifying a Caribbean grouping and a smaller clade linking the Cuyan High Andes with the Patagonian Andes. The Páramo province (Northern Andes) is nested within a Caribbean clade that is a sister area to other Central American terranes of the Central American Arc (Fig. 3.2.: white cross extinct, red cross active). They in turn are sister to the Greater Antillean Arc, a series of Late Cretaceous–Paleogene island arcs formed along CLIP's leading edge(s), parts of which were accreted to South America's northwest margin (Fig. 3.2). Northern STZ is the result of impact tectonics of an ocean plateau (CLIP) with a continent (South America). It represents a mixture of Central and Caribbean biotas carried on their respective arcs and now juxtaposed against the Neotropical biota of cratonic South America.

The other resolved clade in the general areagram links the Cuyan High Andean province to the Patagonia terrane. While their Mesozoic–Paleogene tectonic

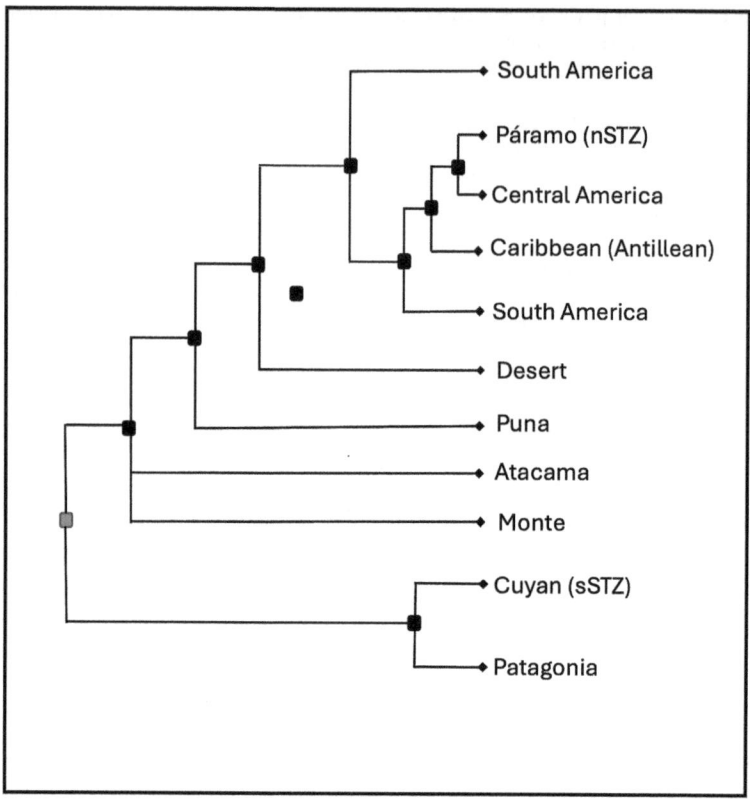

Fig. 3.3 General areagram showing the relationship between areas of endemism in the STZ. Locality of the areas shown in Fig. 3.1 Retention Index (RI) = 0.62, Completeness Index = 24%. RI represents the proportion of the three-item statements that support the intersection tree, and the Completeness Index is a measure of how many three-item statements implicit in the intersection tree are actually present in the original data set. The low value of the Completeness Index is due to the unresolved position of Atacama and Monte in the general areagram

histories are different, they are now stitched together by a Neogene volcanic arc. The sister group relationship shows that the two biotas share a common origin in part and have often been regarded as parts of a larger Austral grouping (e.g. Morrone 2006; Lopez et al. 2008; Roig-Juñent et al. 2018). The southern sector of STZ, particularly the Monte province, is a mix of Neotropical species, derived from cratonic South America, with Patagonia species and palaeoendemic Austral taxa derived from the Patagonia terrane. This part of STZ resulted from terrane accretion and subsequent range expansion and represents a continent-continent transition zone upon which a later ocean-continent tectonic regime was superimposed.

Because the Puna and Desert endemic areas are basal to the Caribbean clade and the areagram has a low Completeness Index (the percentage of three-item statements implicit in the intersection tree that are also present in the original data set), we are wary of over-interpreting the results. While it is possible that their position in the Caribbean clade derives from secondary range expansion southwards, we would like to see this position reconfirmed by phylogenetic analyses of other taxa. The Atacama and Monte provinces are unresolved due to a lack of data and are responsible for this low-value Completeness Index. At a descriptive level, the Monte province shows clear biological links to both Patagonia and cratonic South America (Roig et al. 2009) and may well turn out to be part of the southern transition zone between Patagonian/Austral and Neotropical biotas as suggested by Morrone (2006). The Eastern Cordillera of the Puna province is part of the western margin of the South American craton with its Proterozoic and Palaeozoic basement. The Puna province has been shown to have strong Neotropical links (Urtubey et al. 2010), but future phylogenetic studies may be able to establish other links. The nature of the central part of STZ is unresolved.

3.4.3 Discussion

While we have concentrated on terranes and tectonic history in our review of the STZ, their interaction created real landscapes on which life evolved. How these landscapes (and the way they changed) influenced distributions is key to understanding the development of STZ (Ebach and Michaux 2020). While large-scale tectonic events form the backdrop, the role of dynamic topography can be influential in determining the details. Examples from the Northern Andes, the Altiplano and Cordilleras of the Central Andes, and Patagonia will be discussed to illustrate some of these influences.

While the origin of the North Andes biota can be understood in terms of its tectonic history of terrane amalgamation, its preservation through isolation depended on landscapes that formed barriers to range expansion. A number of studies have suggested that large-scale subsidence during the Paleogene in northwestern South America is a result of dynamic processes (Braun 2010; Shephard et al. 2010; Eakin et al. 2014; Dávila et al. 2019). Shephard et al. (2010) reported dynamic subsidence rates of <40 m/my, which resulted in the creation of an extensive inland basin (~ 106 km^2) of low-lying, swampy wetlands, waterways, and 'islands' of limited relief (the Pebas Wetlands). As Wesselingh and Salo (2006) noted

> 'The Pebas system had a variety of influences over the evolution of Miocene and modern Amazonian biota; it formed a barrier for the exchange of terrestrial biota, a pathway for the transition of marine biota into freshwater Amazonian environments, and formed the stage of remarkable radiations of endemic molluscs and ostracods.'

Antonelli et al. (2009) also discussed the role this wetland system played as a barrier to any floral interchange between the Northern Andes and Neotropics. Lundberg et al. (1998: Fig. 12) showed the Pebas Wetlands and drainage direction at the time of CLIP's collision. It would have been a significant barrier to the mixing of North Andes biota with Neotropical biotas to the east and Central Andean biotas to the south until its disappearance ~ 7 Ma, by which time the North Andes had reached montane heights and a richly endemic biota had developed.

Flat-slab subduction occurs when a subducting slab shallows causing a cessation of magma generation (it is sometimes referred to as a cold subduction) and greater than expected uplift in the overriding plate as cold oceanic crust ploughs underneath. The Altiplano-Cordillera region of the Central Andes is an oceanic-continent complex, with 'normal' subduction building a Late Cretaceous arc inland along the western edge of the South American Craton (Runyon et al. 2022). The continental margin would have had a low relief plain as a result of the dynamics between the underlying denser flat slab and the overlying continental rocks. We assume that the Cretaceous–Paleogene biota was dominated by Neotropical species derived from cratonic South America as there is no evidence of any exotic terranes in this section of the Andes. By the Paleogene, the Western Cordillera had reached heights of 3–4000 m, a result of both volcanism and compression. Uplift and volcanism propagated eastwards across the Altiplano to form the Eastern Cordillera. Uplift was rapid, raising the Altiplano to over 4000 m and the peaks of the Cordilleras to 6000 m creating this high, dry, and exposed environment. Ramos and Folguera (2009) and Runyon et al. (2022) interpreted this rapid rise, extreme crustal thickening, and cessation of arc magmatism to a period of flat-slab subduction during the late Paleogene to early Neogene (c 35–20 Ma). The rapid rise of this montane barrier would have inhibited the east-west range expansion of lowland species and isolated the coastal deserts, promoting the development of endemic biotas. We do not have sufficiently robust data to evaluate a possible Neogene range expansion of species southwards along the Andes Ranges.

The Patagonia terrane is perhaps the most enigmatic part of South America, with little consensus about where it originated, how far it may have travelled, and when it docked (Rapalini et al. 2010). In part this is due to the region's remoteness, poor exposure of the basement due to vegetation cover, and lack of seismic sections due to the area's low oil prospectivity (Ramos et al. 2014). The consensus view is that the terrane is not autochthonous and has come from elsewhere. Oriolo et al. (2023) placed the Patagonia terrane within the Australides orogenic belt in a para-autochthonous position prior to Gondwana's breakup. A close relationship with the Antarctic Peninsula terrane has been suggested by a number of authors (e.g. Poblete et al. 2016; Lagabrielle et al. 2009; Riley et al. 2023), with others suggesting that it is adjacent to the Transantarctic mountains of East Antarctica (e.g. Ramos and Naipauer 2014) or adjacent to Africa (e.g. Ramos et al. 2014; Mundl et al. 2015). Docking time is usually given as late Palaeozoic (e.g. Rapella and Pankhurst 2020), timing at odds with the biogeographic evidence (see below), although Lagabrielle et al. (2009: Fig. 7) hinted at a Neogene suturing. Even the extent of the terrane is debated with some (e.g. Ramos 2008) suggesting that the Cuyan province is part of the Patagonian terrane.

The phylogenetic evidence of the modern biota clearly links Patagonia to other Austral landmasses (Australia) and West Antarctic terranes (Campbell Plateau/ South Zealandia, Tasmania, and the Antarctic Peninsula), and if this linkage dates from the Permian, it would not be detectable in today's biota. Paleogene and Neogene environmental changes in Patagonia have been described by Palazzesi et al. (2021). Earlier Paleogene terrestrial assemblages are diverse (Quattrocchio et al. 2011) and show a strong linkage to Australia (e.g. *Eucalyptus*, casuarinas, and monotremes), but later Paleogene cooling saw the expansion of southern beech and podocarp gallery forest. Towards the end of the Paleogene and at the start of the Neogene, specialised coastal habitats, such as salt marsh, emerged. A significant influx of Neotropical taxa occurred during an early Neogene warm period, with palms and platyrrhine monkeys present (Kay et al. 2021), but these did not survive subsequent cooling and aridification, when today's more open grassland habitats (steppe) were established. On the basis of these data, it would appear that biological evidence supports docking at the Paleogene/Neogene boundary (~30 Ma), although it could be argued that the expansion of Neotropical taxa was promoted by a warming climate and that docking was earlier.

3.5 Conclusions

The major conclusion of our study is that the STZ is non-monophyletic and composite. Rather than a single STZ, there are three distinct zones within it, as it presently understood, only two of which are definitively transition zones, that is, contain a biota that shows two or more phylogenetically distinct groupings. In the north, the zone is transitional in both biology and geology because it was formed when exotic arc fragments carrying a Caribbean biota were sutured to northwest South America. Geological evidence of an extensive Paleogene inland sea and its replacement by a large inland wetland system ensured this biota was largely isolated from its Neotropical hinterland. The breaking of this barrier in the Neogene by dynamic topographic changes to the drainage system that promoted range expansion and the formation of a biologically composite Caribbean/Neotropical biota.

The analysis of the central section of the STZ is inconclusive and requires more detailed research to resolve, but it appears to be a zone of endemism only, with little evidence of a phylogenetically mixed biota. Rather, the rapid uplift of a Neotropical biota from late in the Paleocene created a high altitude, arid, and exposed environment leading to the equally rapid evolution of a unique and endemic biota that contrasts with the adjacent Neotropical humid lowland biota. How much mixing of North Andean, Patagonian, or even North American species has occurred with range expansion along the Andes Range is not clear.

The third, southern transition zone was formed when an Austral biota, containing many palaeoendemic and Austral taxa, was juxtaposed against a Neotropical biota from cratonic South America. The earliest palaeontological evidence for range expansion onto the Patagonia terrane is in the early Neogene, although it cannot be discounted that this barrier-breaking was a consequence of warming rather than docking. If the Cuyan province turns out to be part of the Patagonia terrane, the

Monte province would mark the transition zone. Again, future research will be needed to answer these questions. Like the northern transition zone, the southern sector is both geologically and biologically composite.

References

Allmendinger RW, Jordan TE, Kay SM, Isacks BL (1997) The evolution of the Altiplano-Puna plateau of the Central Andes. Annu Rev Earth Planet Sci 25:139–174

Antonelli A, Nylander JAA, Persson C, Sanmartín I (2009) Tracing the impact of the Andean uplift on Neotropical plant evolution. Proc Natl Acad Sci USA 106:9749–9754. https://doi.org/10.1073/pnas.0811421106

Arzamendia Y, Vilá B (2015) Vicugna habitat use and interactions with domestic ungulates in Jujuy, Northwest Argentina. Mammalia 79:267–278. https://doi.org/10.1515/mammalia-2013-0135

Beck SL, Zandt G (2002) The nature of orogenic crust in the Central Andes. J Geophys Res: Solid Earth 107:ESE-7. https://doi.org/10.1029/2000jb000124

Bell CD, Kutschker A, Arroyo MT (2012) Phylogeny and diversification of Valerianaceae (Dipsacales) in the southern Andes. Mol Phylogenet Evol 63:724–737. https://doi.org/10.1016/j.ympev.2012.02.015

Bohm M, Lüth S, Echtler H, Asch G, Bataille K, Bruhn C, Rietbrock A, Wigger P (2002) The southern Andes between 36° and 40°S latitude: seismicity and average seismic velocities. Tectonophysics 356:275–289. https://doi.org/10.1016/S0040-1951(02)00399-2

Braun J (2010) The many surface expressions of mantle dynamics. Nat Geosci 3:825–833. https://doi.org/10.1038/ngeo1020

Cabrera AL (1971) Fitogeografía de la república Argentina. Boletin de la Sociedad Argentina de Botánica 14:1–50

Caffe PJ, Trumbull RB, Coira BL, Romer RL (2002) Petrogenesis of early Neogene magmatism in the northern Puna; implications for magma genesis and crustal processes in the central Andean plateau. J Petrol 43:907–942. https://doi.org/10.1093/petrology/43.5.907

Ceccarelli FS, Ojanguren-Affilastro AA, Ramírez MJ, Ochoa JA, Mattoni CI, Prendini L (2016) Andean uplift drives diversification of the bothriurid scorpion genus Brachistosternus. J Biogeogr 43:1942–1954. https://doi.org/10.1111/jbi.12760

Cembrano J, Lara L (2009) The link between volcanism and tectonics in the southern volcanic zone of the Chilean Andes: a review. Tectonophysics 471:96–113. https://doi.org/10.1016/j.tecto.2009.02.038

Coira B, Kay SM, Viramonte J (1993) Upper Cenozoic magmatic evolution of the Argentine Puna—a model for changing subduction geometry. Int Geol Rev 35:677–720. https://doi.org/10.1080/00206819309465552

Dávila FM, Avila P, Martina F (2019) Relative contributions of tectonics and dynamic topography to the Mesozoic-Cenozoic subsidence of southern Patagonia. J S Am Earth Sci 93:412–423. https://doi.org/10.1016/j.jsames.2019.05.010

De Silva SL (1989) Altiplano-Puna volcanic complex of the Central Andes. Geology 17:1102–1106. https://doi.org/10.1130/0091-7613(1989)017<1102:apvcot>2.3.co;2

Eakin CM, Lithgow-Bertelloni C, Dávila F (2014) Influence of Peruvian flat-subduction dynamics on the evolution of western Amazonia. Earth Planet Sci Lett 404:250–260. https://doi.org/10.1016/j.epsl.2014.07.027

Ebach MC, Michaux B (2020) Biotectonics: tectonics as the driver of bioregionalisation. Springer. https://doi.org/10.1007/978-3-030-51773-1

Elías GDV, Aagesen L (2016) Areas of vascular plants endemism in the Monte desert (Argentina)

Ezcurra C, Gavini SS, Goldstein M, DellaSala D (2020) Alpine plant diversity in temperate mountains of South America. In: Encyclopedia of the World's Biomes, vol 1, pp 324–334. https://doi.org/10.1016/b978-0-12-409548-9.11906-2

Ferretti NE, González A, Pérez Miles F (2012) Historical biogeography of the genus *Cyriocosmus* (Araneae: Theraphosidae) in the Neotropics according to an event-based method and spatial analysis of vicariance. Zool Stud 51:526–535

Gansser A (1973) Facts and theories on the Andes: twenty-sixth William smith lecture. J Geol Soc 129:93–131. https://doi.org/10.1144/gsjgs.129.2.0093

Garzione CN, Molnar P, Libarkin JC, MacFadden BJ (2006) Rapid late Miocene rise of the Bolivian Altiplano: evidence for removal of mantle lithosphere. Earth Planet Sci Lett 241:543–556. https://doi.org/10.1016/j.epsl.2005.11.026

Giambiagi L, Mescua J, Bechis F, Martínez A, Folguera A (2011) Pre-Andean deformation of the Precordillera southern sector, southern Central Andes. Geosphere 7:219–239. https://doi.org/10.1130/ges00572.1

Goleniowski ME, Bongiovanni GA, Palacio L, Nuñez CO, Cantero JJ (2006) Medicinal plants from the "sierra de Comechingones", Argentina. J Ethnopharmacol 107:324–341. https://doi.org/10.1016/j.jep.2006.07.026

Götze HJ, Krause S (2002) The central Andean gravity high, a relic of an old subduction complex? J S Am Earth Sci 14:799–811. https://doi.org/10.1016/s0895-9811(01)00077-3

Gubbels TL, Isacks BL, Farrar E (1993) High-level surfaces, plateau uplift, and foreland development, Bolivian Central Andes. Geology 21:695–698. https://doi.org/10.1130/0091-7613(1993)021<0695:hlspua>2.3.co;2

Gutiérrez Morales N, Toscano De Brito ALV, Silvério Righetto Mauad AV, De Camargo SE (2020) Molecular phylogeny and biogeography of *Pabstiella* (Pleurothallidinae: Orchidaceae) highlight the importance of the Atlantic rainforest for speciation in the genus. Bot J Linn Soc 195:568–587. https://doi.org/10.1093/botlinnean/boaa092

Hughes C, Eastwood R (2006) Island radiation on a continental scale: exceptional rates of plant diversification after uplift of the Andes. Proc Natl Acad Sci 103:10334–10339. https://doi.org/10.1073/pnas.0601928103

Isacks BL (1988) Uplift of the central Andean plateau and bending of the Bolivian orocline. J Geophys Res Solid Earth 93:3211–3231

Jara-Arancio P, Vidal PM, Panero JL, Marticorena A, Arancio G, Arroyo MT (2017) Phylogenetic reconstruction of the south American genus *Leucheria* lag.(Asteraceae, Nassauvieae) based on nuclear and chloroplast DNA sequences. Plant Syst Evol 303:221–232. https://doi.org/10.1007/s00606-016-1366-7

Kay SM, Maksaev V, Moscoso R, Mpodozis C, Nasi C (1987) Probing the evolving Andean lithosphere: mid-late tertiary magmatism in Chile (29°–30° 30′ S) over the modern zone of subhorizontal subduction. J Geophys Res Solid Earth 92:6173–6189. https://doi.org/10.1029/jb092ib07p06173

Kay SM, Maksaev V, Moscoso R, Mpodozis C, Nasi C, Gordillo CE (1988) Tertiary Andean magmatism in Chile and Argentina between 28 S and 33 S: correlation of magmatic chemistry with a changing Benioff zone. J S Am Earth Sci 1:21–38. https://doi.org/10.1016/0895-9811(88)90013-2

Kay SM, Mpodozis C, Coira B (1999) Neogene magmatism, tectonism, and mineral deposits of the Central Andes (22 to 33 S latitude). In: Geology and Ore deposits of the central, Society of Economic Geologists Special Publication 7, pp 27–59. https://doi.org/10.5382/sp.07.02

Kay SM, Jones HA, Kay RW (2013) Origin of tertiary to recent EM-and subduction-like chemical and isotopic signatures in Auca Mahuida region (37–38 S) and other Patagonian plateau lavas. Contrib Mineral Petrol 166:165–192. https://doi.org/10.1007/s00410-013-0870-9

Kay RF, Vizcaíno SF, Bargo MS, Spradley JP, Cuitiño JI (2021) Paleoenvironments and paleoecology of the Santa Cruz Formation (early-middle Miocene) along the Río Santa Cruz, Patagonia (Argentina). J S Am Earth Sci 109:103296

Lagabrielle Y, Godderis Y, Donnadieu Y, Malavieille J, Suarez M (2009) The tectonic history of Drake Passage and its possible impacts on global climate. Earth Planet Sci Lett 279:197–211. https://doi.org/10.1016/j.epsl.2008.12.037

López HL, Menni RC, Donato M, Miquelarena AM (2008) Biogeographical revision of Argentina (Andean and Neotropical regions): an analysis using freshwater fishes. J Biogeogr 35:1564–1579. https://doi.org/10.1111/j.1365-2699.2008.01904.x

Lundberg JG, Marshall LG, Guerrero J, Horton B, Malabarba MC, Wesselingh F (1998) The stage for Neotropical fish diversification: a history of tropical south American rivers. In: Phylogeny and classification of Neotropical fishes, vol 27, pp 13–48

Martínez GA, Arana MD, Oggero AJ, Natale ES (2017) Biogeographical relationships and new regionalisation of high-altitude grasslands and woodlands of the central Pampean ranges (Argentina), based on vascular plants and vertebrates. Aust Syst Bot 29:473–488. https://doi.org/10.1071/sb16046

Meerow AW, Gardner EM, Nakamura K (2020) Phylogenomics of the Andean tetraploid clade of the American Amaryllidaceae (subfamily Amaryllidoideae): unlocking a polyploid generic radiation abetted by continental geodynamics. Front Plant Sci 11:582422. https://doi.org/10.3389/fpls.2020.582422

Morrone JJ (2006) Biogeographic areas and transition zones of Latin America and the Caribbean islands based on panbiogeographic and cladistic analyses of the entomofauna. Annu Rev Entomol 51:467–494. https://doi.org/10.1146/annurev.ento.50.071803.130447

Morrone JJ (2014) Biogeographical regionalisation of the Neotropical region. Zootaxa 3782:1–110. https://doi.org/10.11646/zootaxa.3782.1.1

Morrone JJ (2015) Biogeographical regionalisation of the world: a reappraisal. Aust Syst Bot 28:81–90. https://doi.org/10.1071/sb14042

Morrone JJ (2017) Neotropical biogeography: regionalization and evolution. CRC Press, Boca Raton. https://doi.org/10.1201/b21824

Morrone JJ (2018) Evolutionary biogeography of the Andean region. CRC Press, Boca Raton. https://doi.org/10.1201/9780429486081

Morrone JJ, Ezcurra C (2016) On the Prepuna biogeographic province: a nomenclatural clarification. Zootaxa 4132:287–289. https://doi.org/10.11646/zootaxa.4132.2.11

Mpodozis C, Ramos V (1990) The Andes of Chile and Argentina. Circum-pacific council for energy and mineral resources. Earth Sci Ser 11:59–90

Mundl A, Ntaflos T, Ackerman L, Bizimis M, Bjerg EA, Hauzenberger CA (2015) Mesoproterozoic and Paleoproterozoic subcontinental lithospheric mantle domains beneath southern Patagonia: Isotopic evidence for its connection to Africa and Antarctica. Geology 43:39–42

Murphy DJ, Ebach MC, Miller JT, Laffan SW, Cassis G, Ung V, Knerr N, Tursky MT (2019) Dophytogeographic patterns reveal biomes or biotic regions? Cladistics 45:4831–4817. https://doi.org/10.1111/cla.12381

Oriolo S, González PD, Renda EM, Basei MA, Otamendi J, Cordenons P, Marcos P, Yoya MB, Justiniano CAB, Suárez R (2023) Linking accretionary orogens with continental crustal growth and stabilization: lessons from Patagonia. Gondwana Res 121:368–382. https://doi.org/10.1016/j.gr.2023.05.011

Padró A, Hechem V, Morrone JJ (2020) Biogeographic characterisation of the austral high Andean district, Patagonian province, based on vascular plant taxa. Aust Syst Bot 33:174–190. https://doi.org/10.1071/sb19005

Parenti L, Ebach M (2009) Comparative biogeography: discovering and classifying biogeographical patterns of a dynamic earth, vol 2. Univ of California Press. https://doi.org/10.1525/california/9780520259454.001.0001

Palazzesi L, Vizcaíno SF, Barreda VD, Cuitiño JI, del Río CJ, Goin F, González-Estebenet MS, Guler MV, Gandolfo MA, Kay R, Parras A, Reguero MA, del Zamaloa MC (2021) Reconstructing Cenozoic Patagonian biotas using multi-proxy fossil records. J S Am Earth Sci 112:103513. https://doi.org/10.1016/j.jsames.2021.103513

Poblete F, Roperch P, Arriagada C, Ruffet G, de Arellano CR, Hervé F, Poujol M (2016) Late cretaceous–early Eocene counterclockwise rotation of the Fueguian Andes and evolution of the Patagonia–Antarctic peninsula system. Tectonophysics 668:15–34. https://doi.org/10.1016/j.tecto.2015.11.025

Pons M, Piceda RC, Sobolev SV, Scheck-Wenderoth M, Strecker MR (2023) Localization of deformation in a non-collisional subduction orogen: the roles of dip geometry and plate strength on the evolution of the broken Andean foreland, sierras Pampeanas, Argentina. Tectonics 42:e2023TC007765. https://doi.org/10.1029/2023tc007765

Quade J, Dettinger MP, Carrapa B, DeCelles P, Murray KE, Huntington KW, Cartwright A, Canavan RR, Gehrels G, Clementz M (2015) The growth of the Central Andes, 22 S–26 S. In: PG DC, Ducea MN, Carrapa B, Kapp PA (eds) Geodynamics of a cordilleran orogenic system: the Central Andes of Argentina and northern Chile, vol 212. Geological Society of America Memoir, pp 277–308. https://doi.org/10.1130/2015.1212(15)

Quattrocchio ME, Volkheimer W, Borromei AM, Martinez MA (2011) Changes of the palynobiotas in the Mesozoic and Cenozoic of Patagonia: a review. Biol J Linn Soc 103:380–396. https://doi.org/10.1111/j.1095-8312.2011.01652.x

Ramirez O, Arana M, Bazán E, Ramirez A, Cano A (2007) Assemblages of bird and mammal communities in two major ecological units of the Andean highland plateau of southern Peru. Ecologia Aplicada 6:139–148. https://doi.org/10.21704/rea.v6i1-2.350

Ramos VA (2008) Patagonia: a Paleozoic continent adrift? J S Am Earth Sci 26:235–251. https://doi.org/10.1016/j.jsames.2008.06.002

Ramos VA (2009) Anatomy and global context of the Andes: Main geologic features and the Andean orogenic cycle. Mem Geol Soc Am 204:31–65. https://doi.org/10.1130/2009.1204(02)

Ramos VA, Folguera A (2005) Tectonic evolution of the Andes of Neuquén: constraints derived from the magmatic arc and foreland deformation. Geol Soc Lond, Spec Publ 252:15–35. https://doi.org/10.1144/gsl.sp.2005.252.01.02

Ramos VA, Folguera A (2009) Andean flat-slab subduction through time. Geol Soc Lond, Spec Publ 327:31–54. https://doi.org/10.1144/sp327.3

Ramos VA, Naipauer M (2014) Patagonia: where does it come from? J Iber Geol 40:367–379. https://doi.org/10.5209/rev_jige.2014.v40.n2.45304

Ramos ME, Folguera A, Fennell L, Giménez M, Litvak VD, Dzierma Y, Ramos VA (2014) Tectonic evolution of the north Patagonian Andes from field and gravity data (39–40 S). J S Am Earth Sci 51:59–75. https://doi.org/10.1016/j.jsames.2013.12.010

Rapalini AE, de Luchi MGL, Dopico CM, Klinger FGL, Giménez ME, Martínez P (2010) Did Patagonia collide with Gondwana in the late Paleozoic? Some insights from a multidisciplinary study of magmatic units of the north Patagonian massif. Geol Acta 8:349–371

Rapela CW, Pankhurst RJ (2020) The continental crust of northeastern Patagonia. Ameghiniana 57:480–498. https://doi.org/10.5710/amgh.17.01.2020.3270

Riley TR, Burton-Johnson A, Flowerdew MJ, Poblete F, Castillo P, Hervé F, Leat PT, Millar IL, Bastias J, Whitehouse MJ (2023) Palaeozoic–Early Mesozoic geological history of the Antarctic Peninsula and correlations with Patagonia: kinematic reconstructions of the proto-Pacific margin of Gondwana. Earth Sci Rev 1236:104265. https://doi.org/10.1016/j.earscirev.2022.104265

Roig FA, Roig-Juñent S, Corbalán V (2009) Biogeography of the Monte desert. J Arid Environ 73:164–172. https://doi.org/10.1016/j.jaridenv.2008.07.016

Roig-Juñent SA, Griotti M, Domínguez CM, Agrain FA, Campos-Soldini P, Carrara R, Cheli G, Fernández-Campón F, Flores GE, Katinas L, Muzón JR (2018) The Patagonian steppe biogeographic province: Andean region or south American transition zone? Zool Scr 47:623–629. https://doi.org/10.1111/zsc.12305

Rojo V, Arzamendia Y, Pérez C, Baldo J, Vila BL (2019) Spatial and temporal variation of the vegetation of the semiarid Puna in a pastoral system in the Pozuelos biosphere reserve. Environ Monit Assess 191:635. https://doi.org/10.1007/s10661-019-7803-7

Rundel PW, Dillon MO, Palma B, Mooney HA, Gulmon SL, Ehleringer JR (1991) The phytogeography and ecology of the coastal Atacama and Peruvian deserts. Aliso: J Syst Floristic Botany 13:1–49. https://doi.org/10.5642/aliso.19911301.02

Runyon B, Saylor JE, Horton BK, Reynolds JH, Hampton B (2022) Basin evolution in response to flat-slab subduction in the Altiplano. J Geol Soc 179:jgs2021-003. https://doi.org/10.1144/jgs2021-003

Shephard GE, Müller RD, Liu L, Gurnis M (2010) Miocene drainage reversal of the Amazon River driven by plate–mantle interaction. Nat Geosci 3:870–875. https://doi.org/10.1038/ngeo1017

Spikings R, Cochrane R, Villagomez D, van der Lelij R, Vallejo C, Winkler W, Beate B (2015) The geological history of northwestern South America: from Pangaea to the early collision of the Caribbean large Igneous Province (290–75 ma). Gondwana Res 27:95–139. https://doi.org/10.1016/j.gr.2014.06.004

Urtubey E, Stuessy TF, Tremetsberger K, Morrone JJ (2010) The south American biogeographic transition zone: an analysis from Asteraceae. Taxon 59:505–509. https://doi.org/10.1002/tax.592015

Wardle P, Ezcurra C, Ramírez C, Wagstaff S (2001) Comparison of the flora and vegetation of the southern Andes and New Zealand. N Z J Bot 39:69–108

Wesselingh F, Salo JA (2006) A Miocene perspective on the evolution of the Amazonian biota. Scr Geol 133:439–458

Zandt G, Leidig M, Chmielowski J, Baumont D, Yuan X (2003) Seismic detection and characterization of the Altiplano-Puna magma body, Central Andes. Pure Appl Geophys 160:789–807. https://doi.org/10.1007/pl00012557

Zaragüeta-Bagils RZ, Ung V, Grand A, Vignes-Lebbe R, Cao N, Ducasse J (2012) LisBeth: new cladistics for phylogenetics and biogeography. Comptes Rendus Palevol 11:563–566. https://doi.org/10.1016/j.crpv.2012.07.002

Chapter 4
Synthesis

Abstract In an effort to build on the approach of Ebach and Michaux (Biotectonics, Springer Briefs, Switzerland, https://doi.org/10.1007/978-3-030-51773-1, 2020) to systematise transition zones, our analysis of Neotropical examples has led us to suggest a number of refinements.

- The central South American TZ (sSTZ) most probably represents an endemic area nested within a Neotropical biota rather than a transition zone, although we do not have the data to discount secondary range expansion from the north (see Fig. 3.3). Whatever future studies may show, it must first be demonstrated that a transition zone has a phylogenetically mixed biota, rather than assuming that the juxtaposition of endemic areas equates to a transition zone.
- We introduce a number of subsettings to both IPB and MPB classification in order to capture the complexity of these systems. By doing so, we hope to increase an understanding of the different types of transition zones and how they might be related to both their tectonic settings and to each other.
- Transition zones are not static and change in space and through time. We explore some of the implications of this view of transition zones as dynamic entities. We conclude that given enough time, transition zones may develop into natural (monophyletic) areas.

4.1 Introduction

The difficulties of incorporating transition zones into a natural area taxonomy were discussed in Chap. 1. Ebach and Michaux (2020) attempted to systematise transition zones using tectonic criteria and suggested that Moronne's (2015) five global transition zones formed two groupings—Wallacea + Mexican TZ + South American TZ and the Saharo-Arabian TZ + Chinese TZ. The former grouping was associated with convergent plate margins and resulted from marginal plate biotectonic (MPB) processes, whereas the latter was the result of intra-plate biotectonic processes associated with dynamic topography (IPB). They suggested that a biotic and tectonic (i.e. biotectonic) approach to historical biogeography could give new insights into understanding transition zones in terms of their tectonic history and ecogeographical development.

While this approach represented a start, we suggest some refinements given the results presented here. Firstly, the South American Transition Zone (STZ) is not a single entity and is itself composite, so the question of how the different parts fit into Ebach and Michaux's (2020) scheme naturally arises. Secondly, and more importantly, a static view of both tectonic processes and transition zones fails to incorporate time and space dynamics. The configuration of land and sea in present-day MPB transition zones is continually changing as the collisional orogenies develop and the tectonics adjust.

4.2   Rethinking Transition Zones

Transition zones are formed when physical (i.e. geographical) barriers separating two or more phylogenetically distinct biotas are removed, allowing a proportion of species from each biota to expand their ranges. While geographic barriers are broken physically and the biotas become mixed to a greater or lesser degree, forming a biologically composite zone of mixed taxa (sensu Morrone 2009). The first step, it is important to establish that there is a mixed biota, that is, to show different phylogenetic links to different source areas. Consider an example of a convergent margin between oceanic and continental plates, such as described in the central part of the STZ.

Figure 4.1a represents the tectonic regime operating in the central Andes. Continental plate 2 represents the South American craton carrying a Neotropical biota; the alpine area represents the High Cordilleras/Antiplano and the oceanic plate as the forearc coastal strip. Is the area of overlap—the High Cordilleras/Antiplano—a true transition zone? While a definitive answer awaits future phylogenetic analysis, we have seen little evidence to suggest the presence of a mixed biota. Rather, the contrast between the flora and faunas of the High Cordilleras/Antiplano and the Neotropical biota of cratonic South America may well be a result of the high levels of endemism induced by the rapid rise of the central Andes and formation of unique ecogeographical habitats. We do not discount secondary range expansion along either the Andes Mountain chain or coastal deserts that may further complicate this picture, but the main conclusion—that a convergent plate boundary may not necessarily produce a transition zone—holds true.

What is missing is that a 'pure' oceanic plate would not carry its own unique terrestrial biota. For transition zones to form along convergent plate boundaries, there must also be two (or more) distinct biotas associated with the different geological terranes. This situation is illustrated in Fig. 4.1b and is exemplified by the northern STZ (arc-continent convergence) and southern STZ (continent-continent convergence), respectively. In both cases, distinct biotas (Caribbean/Central American and

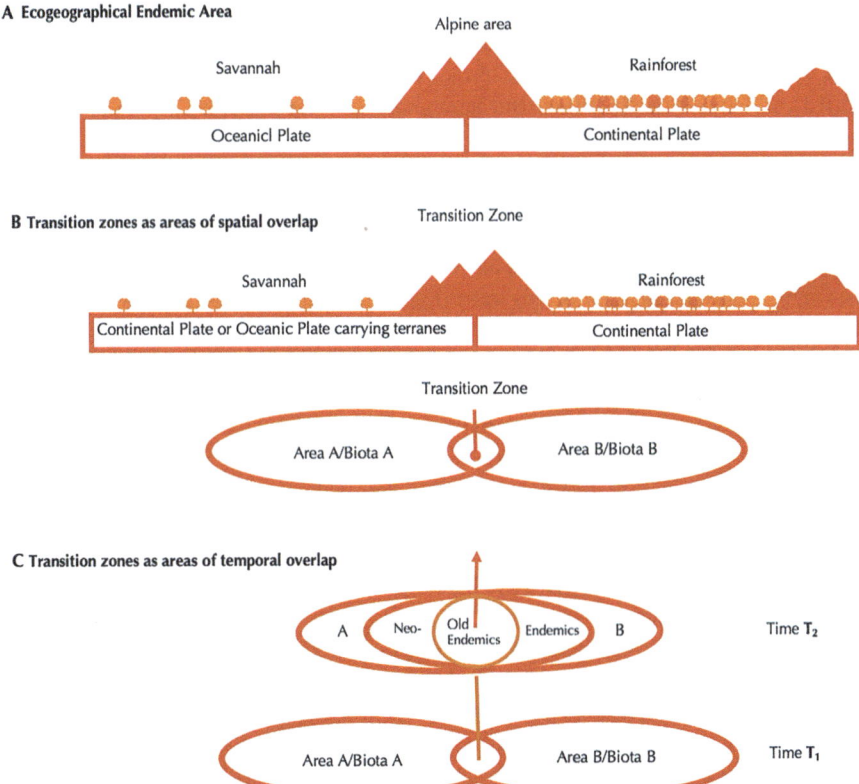

Fig. 4.1 Cartoons showing different ways of viewing transition zones. See text for details (**a**) An example where an ecogeographical zone with distinct endemic areas is formed along a continental arc or intra-plate thrusting via flat-slab subduction without the importation of a phylogenetically distinct biota (**b**) Transition zone as an area of spatial overlap (**c**) Transition zone as an area of temporal overlap

Neotropical; Austral and Neotropical) have been juxtaposed. The process of convergence alters the ecogeographical character of land along the suture zone providing new or novel environments for taxa from both biotas to expand into and exploit. This process need not necessarily be symmetric. In the example illustrated in Fig. 4.1b, taxa already thriving in arid and exposed savannah environments are, we would suggest, more likely to survive in montane environments than rainforest species and, therefore, would be expected to dominate the transition zone biota numerically.

However, the above illustrative examples, however, represent a static view of what is a dynamic process of evolution in space and through time. Consider the process of convergence of two separate continents, each carrying a distinctive biota, can be classified in terms of three phases: rifting, drifting, and amalgamation. In the rifting and drifting phases, an endemic biota develops following vicariance and isolation. The rifting of the South China Block from Gondwana in the mid-Palaeozoic, its northward drift across the palaeo-Tethys, and its eventual amalgamation with a tropical Cathasian biota of the North China Block, described by Metcalfe (2013), illustrates these points. But amalgamation, the final step in the process, is not instantaneous, and the degree of overlap would be expected to increase as the oceanic crust between them is consumed. If we could add a little dynamism to Fig. 4.1b, then the zone of overlap would grow in size as the areas approached each other.

Wallacea (Indo-Malay TZ) represents an early stage in the collision between Asia and Australia. A study of this transition zone (e.g. Michaux 2010; Michaux and Ung 2021) has shown that rift fragments from both continental areas are formed and transported by processes such as back-arc spreading or slab rollback within the collision zone. In addition to these microcontinental blocks, the collision zone is also populated by island arcs that carry their own biotas. The present transition zone in Wallacea is a consequence of this melee of terranes. But what of the future? Extrapolating current plate movements has led some authors (e.g. Safonova and Maruyama (2014)) to suggest that Australia will become sutured to East Asia as part of a new supercontinent Amasia and that the present collision zone will be subsumed within the future suture zone between Australia and East Asia. However, in the case of the Caribbean Transition Zone, the process of amalgamation of South America with North America has been arrested because the thickened LIP of the Caribbean Plate cannot be subducted and destroyed as the cratons approach each other.

Changes in space (geographic size) are inextricably linked to changes through time, and both the extent and characteristics of transition zones are temporally dependent. Figure 4.1c illustrates these points. King and Ebach (2017) demonstrated that Wallacea is 'temporally layered' and that Paleogene and Neogene areas belong to separate clades. Paleogene terranes developed distinctive endemic biotas prior to Neogene additions (mainly from Australasia) forming a mixture of old and new endemics within the modern transition zone. A similar mixture of palaeo- and neo-endemics is also apparent in the sSTZ. At some point in time, a mixed biota will assume an identity separate from the biota surrounding them as levels of endemism increase, and phylogenetic connections become less obvious, for example, through the extinction of source biotas, culminating eventually in the transition zone's apparent monophyly. It is possible that the Chinese and Saharo-African transition zones may have become just this: to appear 'monophyletic' over a period of time due to the extinction of other biota. Until empirical studies of these transition zones are done, the notion that transition zones may become monophyletic over time and thus feature in a hierarchical classification remains just speculation.

4.3   Toward a Bioregionalisation of Transition Zones Using IPB and MPB Subsettings

Table 4.1 sets out a comprehensive range of geological contexts and the possible biological/evolutionary outcomes that may result. Transition zones are formed when phylogenetically distinct biotas are brought together and the barriers previously separating them are removed. In convergent tectonic regimes, this usually involves the destruction of oceanic crust as continents and terranes become amalgamated. Away from active plate margins, barriers can be removed by changes in dynamic topography. When barriers are created, the processes of vicariance (speciation) and endemism are promoted. Our analyses of the Caribbean and South American transition zones have led us to re-evaluate and refine the scheme presented by Ebach and Michaux (2020).

We distinguish two types of intra-plate transition zones, based on their proximity to active tectonic regions and three new subsettings of MPB subsettings of which two can result in the formation of transition zones. Each subsetting reflects differences in the tectonic context, how tectonics creates new topographies and the

Table 4.1 Relationship between tectonic regimes, making and breaking of barriers, endemism, and types of transition zones

	Tectonic regime	Process	Geologically composite	Biologically composite	Outcomes	Example
IPB	Dynamic topography	Breaking barriers	No	Yes	Transition zone	CTZ (c-IPB) SATZ (m-IPB)
		Making barriers	No	No	Vicariance/ endemism	
MPB	Divergent boundary	Making barriers	No	No	Vicariance/ endemism	
	Convergent boundary					
	Ocean-continent	Making barriers	No	No	Vicariance/ endemism	cSTZ
	Continent-continent	Breaking barriers	Yes	Yes	Transition zone	Wallacea (cc-MPB)
						MTZ (cc-MPB)
						sSTZ (cc-MPB)
	Arc-continent	Breaking barriers	Yes	Yes	Transition zone	nSTZ (ia-MPB)
	Multi-plate (impact tectonics)	Breaking barriers	Yes	Yes	Transition zone	Caribbean (ia-MPB)

biological responses to these changing landscapes, and the differences in transition zone characteristics that result. The five IPB and MPB subsettings are as follows:

- Cratonic IPB subsetting (c-IPB): dynamic topography and/or continental tilting induced by distant tectonic activity as in the Chinese Transition Zone (CTZ).
- Marginal IPB subsetting (m-IPB): intra-plate dynamic topography induced along plate margins adjacent to active tectonic zones as in the Saharo-Arabian Transition Zone (SATZ).
- A continental-continental MPB subsetting (cc-MPB): the most common subsetting, which is responsible for the biotic mixing of two continental biotas, as detailed in the discussion of the southern South American Transition Zone (sSTZ; Chap. 3).
- Island Arc MPB subsetting (ia-MPB): an igneous provincial subsetting found along the margins of the Caribbean Plate and in the northern South American Transition Zone (nSTZ).
- Oceanic MPB subsetting (o-MPB): continental arc/plate margin deformation affecting a single biota. Usually resulting in vicariance and endemism but could lead to range expansion along the barrier and the formation of a secondary transition zone (cSTZ). The o-MPB setting is exemplified in the central sector of teh STZ but strickly speaking might not produce what might be termed a primary transition zone.

4.3.1   Cratonic IPB Subsetting (c-IPB)

Definition   Intra-plate stresses caused by distant tectonic activity inducing dynamic topography and/or continental tilt, which may break biogeographical barriers to form composite biotas or create barriers that promote speciation and the evolution of endemic biotas.

Tectonic Regime   Two distinct regimes may occur individually or together, namely, dynamic topography and/or continental tilt. Dynamic topography is caused by processes such convection of mantle material or the presence of fossil subduction slabs within the underlying mantle. Dynamic topography can create depocentres of up to 100–200 m depth over hundreds of square kilometres as the crust responds to downward forces induced by mantle down-welling or increased gravitational attraction due to the presence of cold and dense fossil subduction slabs. In Australia, two such down-wellings of mantle material have been recorded (Sandiford 2007), causing geomorphological anomalies such as river reversals. Similar river reversals, as well as the formation and control of inland seas or extensive wetlands, have been recorded in the Amazon (Shephard et al. 2010) as a result of dynamic topography. The biological consequences of inland seas and the Pebas wetlands on the nSTZ were described in Chap. 3.

Continental tilt due to fast convergent collisions has been reported in the Australian Plate (Sandiford 2007). The tilt has resulted in topographic rises of up to 300 m in the Nullarbor and decreases of 200–300 m along the northern margin of the plate (see Quigley et al. 2010; Ebach and Michaux 2020). Tilting is also a possible cause of drainage flow direction, with much of the central and western parts of the Australian continent draining towards the northeast (Ebach and Michaux 2020). Such tilting may be responsible for the arid nature of the continent's southern margin and interior.

Biotectonic Process and Outcomes Both dynamic topography and tilt have the ability to break barriers. The river reversal in the Amazon may have had a profound impact on biotas due to river capture and the creation of a larger biologically composite area (e.g. Lundberg et al. 1998; Albert et al. 2018). In Australia, continental tilt can affect drainage over vast areas. Without tilt, it is estimated that much of the drainage on the northern margin of the continent would flow inland. Such a change in geomorphology would lead to a savannah interior with permanent lakes, rather than an arid Australian inland (Baker 2023).

Transition Zones The CTZ is classified as c-IPB. Mid-Mesozoic subsidence in eastern China associated with active subduction was reversed ~70 Ma, possibly because subduction moved westwards with respect to the overriding plate, exposing areas of newly exposed coastal plain as the overriding plate rebounded, breaking the former marine barrier that separated the Palearctic and Oriental biotas and promoting range expansion and mixing (Cao 2018; Ebach and Michaux 2020).

4.3.2 Marginal IPB Subsetting (m-IPB)

Definition Intra-plate stresses caused by nearby tectonic activity that induces dynamic topography in response to up- and down-wellings and/or flat-slab subduction.

Tectonic Regime In some ways intermediate between MPB and c-IPB. Dynamic topography induced inboard of active convergence by mantle processes, overriding of subducting plates, or flat-slab subduction.

Transition Zones The Saharo-Arabian TZ, which extends from North Africa and the Arabian Peninsula to Pakistan, contains a mixture of Palearctic and African taxa and is classified as an example of a m-IPB. Convergence between Africa/Arabia and Eurasia started in the Late Cretaceous as Africa advanced northwards, narrowing the intervening neo-Tethys Ocean separating African and Palearctic biotas. Fossil evidence showing the development of a mixed biota (e.g. Kappelman et al. 2003) from the start of the Neogene (Amer and Kumazawa 2005; Pook et al. 2009), possibly earlier in Pakistan (Wüster et al. 2008), is coincidental with a widespread marine transgressions. The cause(s) of the marine transgression is not known but may have been due to a decline in subduction rates under Eurasia and upward flexure of the northern margin of the Afro-Arabian Plate.

If the cSTZ can be shown to have a mixed biota (Austral-Neotropical or Caribbean-Neotopical), then it too would be classified as a m-IPB. In the case of the central Andes, the compressive forces of a convergence zone along the western margin of South America are transmitted far inland by flat-slab subduction, causing rapid elevation of overriding crustal material well away from the zone of active subduction. While mountainous areas form barriers to dispersal across the range, they may promote range expansion along the range.

4.3.3 Continental-Continental MPB Subsetting (cc-MPB)

Definition Convergent continental plate margin collisions that result in the juxtaposition of phylogenetically distinct biotas.

Tectonic Regime Continental-continental collisions can be a complex process that develop through time as continental bocks approach each other via subduction zones and can be arrested if the intervening oceanic crust is part of a LIP (large igneous province), or can be a relatively simple juxtaposition of blocks along transform faults (or highly oblique subduction systems) that cause little compressive deformation.

Biotectonic Process and Outcomes Wallacea represents an early-stage cc-MPB in which a complex melee of rifted microcontinents (from both Sundaland and Australasia) and island arc systems occupy the collision zone. This process of rifting and island arc formation has been occurring episodically since ~45 Ma forming both a spatial and temporal overlap of Asian and Australasian biotas. The Caribbean TZ lies between the South and North American cratons, and its biota is a mixture of Neotropical and Palearctic biotas. The Caribbean TZ was formed towards the end of the Late Cretaceous/Paleogene as the Caribbean Large Igneous Province (CLIP) impacted the proto-Caribbean Ocean/South American/North American system (Fig. 2.2). Arcs and continental fragments were sutured to CLIP's margins as it moved eastwards, colliding with the Bahama Banks in the north and suturing with South America in the south. Because the CLIP is too thick to be subducted, it has effectively stopped the amalgamation of North and South American cratons. The sSTZ represents a cc-MPB formed by a relatively simple juxtaposition of the Patagonia terrane against the South American craton. With little deformation there was probably little creation of new land, but the juxtaposition of different biotas (Austral and Neotropical) resulted in a partial exchange towards the end of the Paleogene (~30Ma).

Transition Zone Wallacea is an early-stage cc-MPB; MTZ, sSTZ and North America/South America represent end-stage cc-MPB.

4.3.4 Island Arc MPB Subsetting (ia-MPB)

Definition Suturing of Island Arcs to continental margins. Arcs may be a single system, complex (an amalgamation of arcs of different ages), or composite (combined with, or built on, continental crust).

Tectonic Regime All modern island arcs carry distinctive biotas and because of their tectonic setting tend to be highly mobile. We see no reason why these characteristics would have applied to past arc systems, based on the uniformitarianism principle that the present is the key to the past. Island arcs always form above subduction zones, either as continental margin arcs or intra-oceanic arcs. Intra-oceanic arcs, i.e. arcs that are presently found within oceans, make up 40% of all modern arcs and are mostly found in the western Pacific. They can be built on oceanic crust formed at mid-ocean ridges (rare in the modern world), in back-arc spreading centres, on oceanic plateaux, or in forearc environments (Leat and Larter 2003) and are characterised by primary melts uncontaminated with continental material.

However, the assumption that an intra-oceanic arc's present position is its defining feature needs to be critically examined in light of its highly mobile nature and the discovery of ancient zircons in many intra-oceanic arcs (Kröner 2010). For example, the modern Solomon Island volcanic arc contain xenocrysts with Archean-aged zircons that imply a continental origin and likely continental material at depth, although secondary contamination following a Paleogene collision with a continental fragment is also a possible explanation (Tapster et al. 2014). As Kröner (2010) noted, the presence of more ancient zircons in intra-oceanic arcs, a seemingly common phenomenon that dates back to at least the Neoproterozoic, poses problems for a simple tectonic interpretation of their origin. We suspect that many intra-oceanic arcs start off life as continental arcs that become detached through processes such as back-arc spreading or slab rollback and end up migrating away from continental margins, where they become buried under effusive arc volcanics. Others may form outboard from continental margins but interact with continental terranes within mobile zones. We have been unable to find any clear explanation or modelling that might explain why a subduction zone should form within what is relatively homogeneous oceanic crust, what might be the origin of crustal weakness along which subduction might be initiated, and what is the source of localised compression in such an environment. Only volcanoes formed at mid-oceanic ridges can be regarded as purely oceanic.

Biotectonic Process and Outcomes Island arcs have been incorporated into continental orogenic belts from at least the Neoproterozoic and appear to be ubiquitous in these tectonic settings (Safonova et al. 2017). The amalgamation of simple, complex, or composite arcs, which may have travelled long distances, can juxtapose phylogenetically distinct biotas creating mixed biotas after amalgamation with continental margins.

Transition Zones A number of ia-MPB transition zones occur in the Caribbean. The nSTZ consists of Late Cretaceous Antillean Arc material and Cretaceous/Paleogene Central American Arc material accreted to the shared continental margin of northwest South America as CLIP's migrated eastwards migration and displacing proto-Caribbean crust. The Greater Antilles/Bahama Banks also represent an example of a composite ia-MPB transition zone because a Late Cretaceous arc was either built on a Grenville-aged continental margin or incorporated such material as CLIP's impacted on rifted continental fragments along the margin of the proto-Caribbean Ocean as detailed in Chap. 2. The collision of this composite arc with the Bahama Banks, a North American fragment, formed a transition zone between the Caribbean (Neotropical) and Palearctic biotas. The modern Lesser Antilles Arc, while seemingly an example of a simple arc system, may well be built on Late Cretaceous arc material. In any event, the biota of the Lesser Antilles appears to be overwhelmingly South American taxa rather than being a transitional.

References

Albert JS, Val P, Hoorn C (2018) The changing course of the Amazon River in the Neogene: center stage for Neotropical diversification. Neotrop Ichthyol 16:e180033. https://doi.org/10.159 0/1982-0224-20180033

Amer SAM, Kumazawa Y (2005) Mitochondrial DNA sequences of the Afro-Arabian spiny-tailed lizards (genus *Uromastyx*; family Agamidae): phylogenetic analyses and evolution of gene arrangements. Biol J Linn Soc 85:247–260. https://doi.org/10.1111/j.1095-8312.2005.00485.x

Baker M (2023) Influence of dynamic topography to deposition and the evolution of the Australian landscape through numerical modelling. PhD Thesis UNSW

Cao K (2018) Cretaceous terrestrial deposits in China. China Geol 1:402–414

Ebach MC, Michaux B (2020) Biotectonics. Springer, Switzerland. https://doi.org/10.1007/978-3-030-51773-1

Kappelman J, Tab Rasmussen D, Sanders WJ, Feseha M, Bown T, Copeland P, Crabaugh J, Fleagle J, Glant M, Gordon A, Jacobs B, Maga M, Muldoon K, Pan A, Pyne L, Richmond B, Ryan T, Seiffert ER, Sen S, Todd L, Winkler A (2003) Oligocene mammals from Ethiopia and faunal exchange between afro-Arabia and Eurasia. Nature 426:549–552. https://doi.org/10.1038/nature02102

King AR, Ebach MC (2017) A novel approach to time-slicing areas within biogeographic-area classifications: Wallacea as an example. Aust Syst Bot 30:495. https://doi.org/10.1071/sb17028

Kröner A (2010) The role of geochronology in understanding continental evolution. Geol Soc Lond, Spec Publ 338:179–196. https://doi.org/10.1144/sp338.9

Leat PT, Larter RD (2003) Intra-oceanic subduction systems: introduction, vol 219. Geological Society, Special Publications, London, pp 1–17. https://doi.org/10.1144/gsl.sp.2003.219.01.01

Lundberg JG, Marshall LG, Guerrero J, Horton B, Malabarba MCS, Wesselingh F (1998) The stage for Neotropical fish diversification: a history of tropical south American rivers. In: Phylogeny and classification of Neotropical fishes, vol 27, pp 13–48. https://doi.org/10.1016/b978-0-12-815872-2.00001-4

Metcalfe I (2013) Gondwana dispersion and Asian accretion: tectonic and palaeogeographic evolution of eastern Tethys. J Asian Earth Sci 66:1–33. https://doi.org/10.1016/j.jseaes.2012.12.020

Michaux B (2010) Biogeology of Wallacea: geotectonic models, areas of endemism, and natural biogeographical units. Biol J Linn Soc 101:193–212. https://doi.org/10.1111/j.1095-8312.2010.01473.x

Michaux B, Ung V (2021) Biotectonics of Sulawesi: principles, methodology, and area relationships. Zootaxa 5068:451–484. https://doi.org/10.11646/zootaxa.5068.4.1

Morrone J (2009) Evolutionary biogeography: an integrative approach with case studies. Columbia University Press, New York

Morrone JJ (2015) Biogeographical regionalisation of the world: a reappraisal. Aust Syst Bot 28:81. https://doi.org/10.1071/sb14042

Pook CE, Joger U, Stümpel N, Wüster W (2009) When continents collide: phylogeny, historical biogeography and systematics of the medically important viper genus *Echis* (Squamata: Serpentes: Viperidae). Mol Phylogenet Evol 53:792–807. https://doi.org/10.1016/j.ympev.2009.08.002

Quigley MC, Clark D, Sandiford M (2010) Tectonic geomorphology of Australia, vol 346. Geological Society, Special Publications, London, pp 243–265. https://doi.org/10.1144/SP346.13

Safonova I, Maruyama S (2014) Asia: a frontier for a future supercontinent Amasia. Int Geol Rev 56:1051–1071. https://doi.org/10.1080/00206814.2014.915586

Safonova I, Kotlyarov A, Krivonogov S (2017) Intra-oceanic arcs of the paleo-Asian Ocean. Tectonophysics 8:547–550. https://doi.org/10.5800/gt-2017-8-3-0287

Sandiford M (2007) The tilting continent: a new constraint on the dynamic topographic field from Australia. Earth Planet Sci Lett 261:152–163. https://doi.org/10.1016/j.epsl.2007.06.023

Shephard GE, Müller RD, Liu L, Gurnis M (2010) Miocene drainage reversal of the Amazon River driven by plate–mantle interaction. Nat Geosci 3:870–875. https://doi.org/10.1038/ngeo1017

Tapster S, Roberts NMW, Petterson MG, Saunders AD, Naden J (2014) From continent to intra-oceanic arc: zircon xenocrysts record the crustal evolution of the Solomon Island arc. Geology 42:1087–1090. https://doi.org/10.1130/g36033.1

Wüster W, Peppin L, Pook CE, Walker DE (2008) A nesting of vipers: phylogeny and historical biogeography of the Viperidae (Squamata: Serpentes). Mol Phylogenet Evol 49:445–459. https://doi.org/10.1016/j.ympev.2008.08.019